*Of a Small and Modest Malignancy,
Wicked and Bristling
with Dots*

Of a Small and Modest Malignancy, Wicked and Bristling with Dots

NORMAN MAILER

1980: LORD JOHN PRESS

NORTHRIDGE, CALIFORNIA

Lord John Press
19073 Los Alimos Street
Northridge, California 91326

ISBN 0-935716-05-x

THIS SUMMER *I took my family on a motor trip through the West and left the highway for a one-hundred-mile jog to pick up some fossils for my boy. This took us to Hurricane, Utah . . . around Zion Park country. A great vastness, magnificent. . . . It is three in the afternoon when we get there, to a house, the fossil seller we read about in the Museum of Natural History magazine operating out of his cellar. It's a gorgeous day, just right, and we go in—into a small living room with about nine people in it, all staring at that square of light: they're watching a game show, three generations of them, the grandparents (American Gothics, blooded, stolid) down to the near infants, wired, a funny show, I guess it's supposed to be, at least "joyous," but no one is laughing, no one is smiling, even; they're just fixed there in their places, staring (not looking, mind you, in the way that Sontag separates looking and staring: by God, this is "silent" art for the populace). Well, the feller sends us down to the cellar to pick around in his stuff and goes back into the living room. We're down there almost three hours; my boy can't make up his mind, and my wife is no help; and when we come back up, they're there, same postures, no change; a soap opera, I think it was, running. . . . It leaves you feeling something about the force we are talking about, its power to denature.*

—out of a letter from Gordon Lish

Of a Small and Modest Malignancy, Wicked and Bristling with Dots

LIMBO was a house of many mansions. Mailer's spirits, while hardly free—they spun around his soul like electrons in a ring—began, nonetheless, to awake. The telling monotonies of Limbo—those stupors that drifted like bad weather, apathies piling on apathies like old newspapers, the crackle of static, the playback of cocktail gabble, and the stations of the cross of feeling empty while waiting for subway trains and airline shuttles and waitresses in busy lunchrooms—having been the only items so far experienced, Mailer began to believe that this enforced immersion in every repellent experience (his vision filled with nothing but the faces of digital watches, the smell of pharmacies, the touch of polyester shirts, the wet wax paper of McDonald's hamburgers, the air of summer traffic jams, and shrieks of stereo as the volume is mislaid, even the little nausea that plastic highball glasses will give to the resonance of booze) was not necessarily going to scourge him around one eternity before dispatching him to another, but might be instead his expiation. In place of his soul expiring altogether, or suffering some holocaust of damnation, he might still belong to his old karmic chain and therefore be going through a purification of misspent hours.

If all who died were guilty in part, at least they were also innocent, and Limbo, thought Mailer, might even be the charity to suppose that one was in need of education. The inhabitants were judged by one fine measure: had they or had they not wasted more of the soul's substance than was required by the difficulties of their life? Since the first information offered here was that the most vulnerable fault of the soul was nothing other than to squander time—which time, whole and mysterious bed of light, electricity, and force, had been invested, like the true fund of the realm, in every soul—it followed that time was not to be wasted, but was rather—whatever foul and fear-filled habits taken into account—to be spent as wittily, cheerfully, and/or bravely as possible. That was the standard of Limbo. Time was not to be wasted.

It was therefore now his to recognize that for every dull thing one had done to the universe (filling out crossword puzzles one did not wish to fill out, taking the Eastern Airlines shuttle when other transportation was available), every hour one had voided thereby of its fresh and keen desire to be used, the air around one, in consequence, suffocated by psychic exhaust, had now to be breathed again. This mansion of Limbo was here to bring you face-to-face with those sins for which there are no tears, even as a husband and wife cannot weep if they lose a potentially heartfelt piece of ass by watching TV all night; yes, this corner of Limbo (a clean and well-appointed place whose appurtenances were suggestive of the interior of a picture tube on a black-and-white set—that is, all curved, silver, and grey, an odorless impalpability of fluorescence in an eternity of flickering) was now apparently ready to teach him, that is, teach his soul, something new. This meant, on reflection, that he might still have a soul. Something in himself exactly like the Old Center seemed certainly not to have ceased, that part at least of himself he had always thought wiser than the rest (that part ready to mash an invisible egg on his head after a dead-ass remark) was coming to recognize that his tour through Limbo (if indeed a tour and not an ongoing punishment) would ask him to meditate at length and presumably to purpose on the yaws and palls of his life that had passed through TV. He was going to be obliged to look back on his wretched collaboration with the multimillion-celled nausea machine, that Christ-killer of the ages—television. (Let us say it takes a Jew not wholly convinced of the divinity of Christ to see that is whom the tube is killing.) And he shuddered through the now familiar, if minimal, retchings of Limbo while remembering how on numerous occasions, with each of his seven children, he had closed the doors of his own resistance to TV and let the little fuckers keep looking at the screen because it *pacified* them, which is to say took the lividity of their five-year-old nerves and slowly (that is, faster than sight) and buzzingly cauterized their nerve ends just the right bloody bit, no blood seen. Again that guilt which cannot be alleviated by tears stirred like sludge in his own small part of the great cosmic gut.

So was he obliged to brood over the nature of television. As soon brood over the nature of cancer gulch! Might they not be the same? There was a malignancy present in the bowels of communication, and it was video.

He prepared therefore for the meditative journey that proves most excruciating in Limbo, a rounding through the past, a trip back! It is a venture full of perils. To meditate on TV might prove equal to writing a recollection of an enemy one has never met and cannot quite believe in. Indeed, how to conceive of an enemy who is without personal animosity? It was like writing a memoir of an oxymoron. Limbo set its tasks.

YES. TASKS. In Limbo, concepts of punishment might as well have been dictated by disciplinarians of the stature of Vince Lombardi and Mao Tse-tung. He would now be obliged to trace the leaching of his soul through fifty—or was it a hundred? or two hundred?—TV appearances (the promiscuous are not those who know too many lovers so much as those who no longer count), and extract from the memory of these outings a way to absolve the excrementitiousness of his TV past.

Yet he could not even know that this was his task. He found, out of the thousand papers he might have managed to smuggle across with him, nothing in Limbo but a letter from Gordon Lish, an editor he hardly knew; on the evidence of this letter, Lish had a style whose carefree disjunctions were much like the author's own, and that was swell. That was company on a long trip.

On the other hand, Mailer had small trust in Limbo. There was no way of knowing the certifiability of what you could remember. Lish's letter seemed to propose an immense essay: *Treat television,* Lish had written, *as pure phenomenon, purest form. . . . Open the eye onto television without comment, without attitude, without the particularities that constitute the content of the thing as it is.* So went a couple of the phrases. To the degree that Mailer thought he understood what this meant, he thought he understood it well: *Television will be displayed wholly in terms of its existence as the all-pervasive environment, the universally felt national experience . . . an awesome, omnipresent force, the object that has penetrated and now fused with its subject . . . a revelation of TV as gestalt, form, power.*

What if he had written the piece already? What if it had been a good piece, and he had only brought Lish's letter along to remind himself of a nice achievement? The most damnable property of Limbo (other than its nearness to damnation itself) was that it made one's memory more appalling than when one had a hangover. It certainly felt as if he *had* written such a piece, or was it only that on every day in the past when he turned on a television set, he had felt the emotions Lish was exhorting him to put upon the page: *the immense circuitry, something is beaming, beaming, beaming.* Yes, he thought of that every time he was near a set. It was a most unpleasant *something.* So unpleasant that over the years he had virtually stopped looking at TV. Even the study of professional football games on Sunday afternoons in a room full of good friends and bloody marys had begun to pall. The company was good, the drinks were fine, but *something* was in the room, and he could feel its presence eating them. Yes, you did not make love to your wife after an evening of TV. *Something* had entered the blood; *something* had been busy eating the sweetmeats out of the seed.

It was true. It was a personal phenomenon. He had stopped watching the tube. He had no interest even in what was declared good. Once he turned on *Mary Hartman, Mary Hartman* and could see why it was being talked about: yes, it was tastefully written, it took chances, was offbeat, but *something* was right in there with it and he didn't like *Mary Hartman* enough to keep the taste of aspirin in his mouth. *Roots* went by; he never took a look. "It's a cultural phenomenon," friends argued with him. He was experiencing his own phenomenon. No more TV. Unlike cigarettes or booze, TV was the easiest habit in the world to give up—once it had leached you out. Every night, or almost every night, even three nights a week, for twenty years, the vampire that lived at the center of *something* had come out of the set and sucked his blood; so it was small surprise that *something* had no interest left in him either, not after twenty years of getting his blood ready for Limbo. Of course, now it was not a problem. There were no TV sets in Limbo. Indeed, how could there be if it was understood that for the duration of one's stay in the mansion one was supposed to reabsorb from the air those vibrations one's set had put into it, no better than an intellectual cow redigesting the waste injected into human stuff, the *small living room with about nine people in it, all staring at that square of light . . . no one is laughing, no one is smiling even; they're just fixed there in their places, staring. . . .* It was part of the malice of Limbo that Mailer now began to brood on the possibility that these people in the little Utah ranch house were looking at him, were stupefiedly studying his own electronicized limbs and lineaments in a replay of some godawful afternoon show ten years old—what else would they choose to replay for these black-and-white sets in the desert? "Norman," said Joe Franklin, beaming in the midday sun of the TV lights, "you're a great writer, I hear—he's a great writer, folks, he writes books!—and I don't read books much, don't have the time, so tell me, what's the name of a good book?" Into the silence, Joe says—he has a pipeline into *something*—Joe asks: "Is *Alice in Wonderland* a good book?"

It cannot be a put-on. Franklin's eyes bounce like marbles.

"Yeah," says Norman Mailer, "it is a good book."

"How about *Rebecca of Sunnybrook Farm,* Norm?"

"I would guess that's not as good." When this is over, Mailer tells himself, he will kill the publicity girl from New American Library who put him on this show in the first place. It will sell books, she had promised him.

Out in Utah, are they taking it all in?

Is he more guilty for having made personal appearances on TV shows than those who only refuse to turn off their sets?

In the years before he showed up on TV, he used to watch it religiously; after a while, sacrilegiously. It started the first winter he smoked marijuana. He smoked it with all the seriousness of what was then his profoundly serious heart. It was 1954, and the drug was more important than any love affair he had had. It taught him more. Making love to different women, he would attempt to find the place where marijuana had last left him. It was the arena of the particular sensation he chased, as though he had been given a lovely if ineluctable emotion while watching a bullfight, and so he went back to the plaza to look for the same emotion —it did not matter who the bullfighter was.

Since he was then in the opening out of a career that would later provide a false legend of much machismo, he was still timid. Being deep in pot, and relatively full of life, certainly full of every intimation about himself, he was nonetheless too timid to go out late at night and see what the bars would provide. Since his second wife, not unlike himself in her jangled relations to bravery and cowardice, had usually, and prudently, gone to bed, he would be up alone at night, his mind teeming, and he would watch TV till the stations shut down. In those days, he made monumental connections on pot. He had to see no more than one animated spiral released by a commercial into the guts of a washing machine, a lively spiral of light that would tunnel right up the tube, and he would try to explain to his friends the next day that the advertising agency was promoting the idea that their washing machine was congenial to a housewife's cunt.

His friends thought him mad. He examined automobile commercials in the same light and saw that they were no longer selling the car by way of the pretty girl sitting on the fender, as once they did; now, they were selling the car itself. The car was the fuck. "Dynaflow does it in oil," the announcer would say of an automatic transmission. So he would tell his friends. His friends would think him mad and try to dissuade him from smoking too much pot. Marijuana was regarded differently in those days; it offered echoes of *Reefer Madness.*

He would watch Ernie Kovacs and Steve Allen late at night and recognize that they knew what he knew. They saw how the spiral worked in the washing-machine commercial and why Dynaflow did it in oil. Years later, when motivational research was presented to the world, and everybody was ready to tell everybody at a party that an automobile was not used by a man to get a mistress but was the mistress, and that a housewife liked to identify the health of her washing machine with her own genitourinary harmonies—speak no ill of the bowels!—Mailer was merely glad that Vance Packard had done the job. He was behind on too many of his own jobs. Marijuana had flung the separate parts of his brain into too many vivid places. In those days he had perceptions on every subject; was convinced, as a consequence, that he was a genius. He was getting very little writing done.

Still, he clung to his set. It explained the world to him. He was getting hip to everything, and the beauty was that he did not have to venture out.

It would not occur to him until six years later, when

he stabbed his wife, that it was not timidity that had been his first vice, but violence, a murderous nest of feeling so intransigent that he did not dare to go out at night, for good cause, and did not know how to sleep at night without Seconal, for better cause—there was too much hatred of the distance between what he wished to do and what he was able to accomplish. Since his wife, faced with the choice of going to sleep early or entering a claustrophobic quarrel, would, of course, go to sleep, he would sit by himself from midnight until two in the morning, when the last show would go off the air, the flag would ripple in the wind, and *The Star-Spangled Banner* would be played. In those days, he got to hate *The Star-Spangled Banner* all over again. It sounded like the first martial strains of that cancer he was convinced was coming on him. And who knows? If he had not stabbed his wife, he might have been dead in a few years himself—the horror of violence is its unspoken logic.

So, in those early mornings when television was his only friend, he knew already that he detested his habit. There was not enough to learn from watching TV. Some indispensable pieces of experience were missing. Except it was worse than that. Something not in existence was also present, some malignancy to burn against his own malignancy, some onslaught of dots into the full pressure of his own strangled vision. Often, when the stations went off the air and no programs were left to watch, he would still leave the set on. The audio would hum in a tuneless pullulation, and the dots would hiss in an agitation of unlabeled forces. The hiss and the hum would fill the room and then his ears. There was, of course, no clamor—it was nearer to anti-noise dancing in eternity with noise. Yet watching the empty video, he would recognize it was hardly empty. Bands of grey and lighter grey swam across the set, roll-overs swept the dots, and something like sunspots crackled forth. Then the set went back to slow waves and the drone of the audio. He discovered at last that such use of TV was a tranquilizer to deaden the sharpest edge of his nerve. Palpably blunted by a half hour of such odorless immersion, he would, with the aid of his Seconal, be a little more ready to go to sleep.

A few years later, when McLuhan would torment the vitals of a generation of American intellectuals with the unremovable harpoon that "the medium is the message," Mailer could give his agreement. The message of TV was the scan of grey on grey and the hum of the sound when there is neither music nor voice. Much later, in the fall of '72, he would set out to make audiences laugh by comparing President Nixon's victorious but still featureless personality to a TV screen that is lit when nothing is on the air. Nixon was there, he would remark, to deaden the murderous mood of the Republic. Indeed, it was the best explanation for why a man thus unpopular would win by so great a majority. If Nixon did not make anyone happy, neither did the TV set. Its message was equal to Nixon's: I am here to deaden you —you need it!

Maybe the U.S. did. Brooding over America of the Sixties, that insane, expanding America where undercover F.B.I. men were inspiring (wherever they were not committing) the most violent acts of the left, brooding on that rich and powerful country whose Puritanism was still as alive as every Baptist; that corporate land with no instinctive response to aesthetics engaged now in the dissemination throughout the world of the worst applied aesthetics in the history of the world, its superhighways the highest form of strip-mining, its little buildings kin to shoe boxes, its big buildings scaled to the module of one cardboard carton piled on top of another, and another, the U.S. skyline thereby deserting the high needles of Manhattan for the Kleenex boxes of Dallas; that food-guzzling Republic which froze its food before it overcooked it, and liked to lick ketchup off french fries so soggy they drooped from your fingers like worms (the worst food in the history of the world!); that sex-revolutionary Republic where swinging singles connected up with like units (every other Baptist!); that sadistically revolutionary Republic going into black leather, S & M, and gooks defoliated with the foliage; that young Republic digging the hog resonance of motors between one's legs, and flamethrowers and comic books, and Haight-Ashbury and hitting golf balls on the moon; yes, so brooding he came to decide that America, full of dread, could certainly use TV: I am here to deaden you—you need it!

Those were legitimate feelings for 1972. One had spent by then a large fraction of one's life watching TV, and one had long ago had the education of putting oneself on shows. Back in '54 and '55, however, stoned on pot, wafted up and down on Seconal, jammed with ambition, terror, and the common lust to learn the secrets of the world some easy way, the immersion in

TV was profound. By it one could study the world, and the tricks of the world.

So, for instance, in 1955, would Mailer examine people as strange to him as Igor Cassini, for the columnist had a show in those years for snobs, and it was hard as fiber glass. Of course, TV was more interesting then. You could see a genuine article regularly—well-groomed, Republican, rich enough to own horses, sexy enough to marry up, and empty enough to find any topic of conversation amusing, provided it was void of content. Those were Igor Cassini's guests. Mailer would on such occasions peer right into the tube to get a little nearer to Fitzgerald's wealth.

Or: studying the American tourist, he learned much about fellatio in the Eisenhower era. TV was scintillating for that. Next to the oil of Dynaflow and the spiral in the washing machine came the phallic immanence of the microphone. A twinkle would light up in Steve Allen's eye as he trundled his mike and cord down the aisle and in and out of impromptu interviews with his audience, parking the rounded knob right up against the mouth of some starched skinny middle-western matron, lean as whipcord, tense as rectitude, a life of iron disciplines in the vertical wrinkles of the upper lip; the lady would bare her teeth in a snarl and show a shark's bite as she brought her jaws around to that black dob of a knob so near to touching her tongue.

A high-school girl would be there with the graduating class on a trip to New York. She would swoon before the mike. She would not be able to get her mouth open. She would keep dodging in her seat, and Steve would stay in pursuit, mike extended. Two nights before she had dodged for hours in the back of a car. My God, this was in public!

A young housewife, liberal, sophisticated, happy to present her congenial point of view, now compliments Mr. Allen on the quality of his show. "We watch you regularly, Steve, and like to think we're not too far behind the times up in Norfolk, Connecticut. That's right, Norfolk, not Norwalk." Her mouth, which has regular lips, is held a regular distance from the microphone. She shows no difficulty with it, no more than she would have with a phallus; two fingers and a thumb keep the thing canted right. There can be nothing wrong, after all, in relations between consenting adults. So speaks her calm.

Then there is a big heavyset man who owns a grain-and-feed store in Ohio. He prides himself on imperturbable phlegm and some thrift with words. He is not quite aware of the mike. If a man came into his store and proceeded to expose himself, this proprietor would not see it right away. He might, after all, be comparing the merits and demerits of one feed mix with those of another; just as he chews no gum when he walks, neither does he offer attention to tangentia as he talks. Now, singled out from the audience to be interviewed, he is stiff-necked, and responds only from that side of his mouth adjacent to the cheek to which Steve is asking the questions; he allows, "New York is a good place to visit, but, yessir, I'll be glad to start up for home," then, *bla-looh!* he sees it, black blimp-like little object! He blinks, he swallows, he looks at Steve: "I guess I've had my say, Mr. Allen," he says, and shuts up shop. Later he will tell a pinochle partner about the crazy people in New York.

"Steve," says the next fellow, "I'm awfully glad you selected me. I've always wanted to talk to you." He is fully aware of the mike and what it portends. "Yes, yes, I'm a male secretary, love the work." "It doesn't bother you," asks Steve, "if people say, 'What is that, a male secretary? Isn't it supposed to be women's work?' " "Oh, Steve, that doesn't bother me a bit. Here," he says, reaching for the mike. "Do you mind? I'm much more comfortable when *I* hold it."

"Help yourself," says Steve.

"Oh, I intend to," says the guest. "Life is a feast, and I think we should all get what we can, don't you?"

Hip! Mailer, building dreams of future power in the darkness, saw it all. America was getting hip. In the center of marijuana was the secret of how to do it. You needed the audacity to take your step.

IN THOSE DAYS his life was simple, if much debilitated. Exhausted from writing *The Deer Park,* frantic from trying to do his columns for *The Village Voice,* used up from the terror of his central political premise—he believed that the orgy was the salvation of America (he still did not know that the average swinger was about as militant as the average tennis player, and in an equal rush to save the cosmos)—he was, if these three times exhausted, nonetheless a demon of psychic r.p.m.'s. When his brain was not in torpor, it felt like an immense dynamo running at top speed in an empty shed in the middle of the woods. He knew what it was to feel like electricity, and understood the profoundly depressing statement in *Being and Nothingness* that consciousness was the nihilation of Being. For his brain worked at a great rate (half as fast, it must be on occasion, as Jean-Paul Sartre's), and he felt the full dimensions of the void his consciousness was generating in his brain cage.

Still, he was also possessed of an idea. It would go through many changes and verge on becoming the opposite of itself, but it would remain the idea of his life—it was (to take it by the most dangerous premise first) that criminals are nearer to God than cops; corollary: A good cop is therefore a work of art. It would also conclude that love is as rich a mystery as death, and perhaps as indecipherable; you could hardly know more than whether you had a love that was good or bad for your orgasm. In consequence, the idea would state that fine justice could be found only once in one's life, and that was in the hour after death; something quite the opposite of *something* looked at you from the other side.

He had the confidence that ideas such as these would eventually thrive unless the liveliest nerve of the mind was doomed. If such doom was a real possibility, and the totalitarianism of the twentieth century was not the shadow but the reality, and one lived beneath it, well, society was not yet dead, and he still saw a great mission for himself: it was to save the nerve of Being.

Since he had also to recognize that nine tenths of what went on in television looked to kill the nerve, his sense of mission grew larger. It was the backstretch of the Eisenhower years. There was not much hope on any horizon. Not a clue on how to stir. Since he did not know a revolution was beginning in America that would live for a little while on an exchange of consciousness between blacks and whites, and a beat generation would yet poke holes in the American middle class, and a civil-rights movement shift the nature of the laws, he thought in the measureless vales of his vanity that he was obliged to do the job himself, and he had mental experiences out on marijuana so harsh, so arrogant, so lonely, so tormented, and so demon-ridden—as if twenty devils could pull his chariot when ordered—that he was left with the confidence for years to come that he could write about the inner states of men like Hitler and Napoleon, Lenin and Castro and Cortés. It is a vanity to keep a young writer alive.

Of course, it also made him look on every action as immense with consequence—the early steps of a great leader! He did not forgive friends lightly for failing to see the transmogrificational significance of his thought, nor would he forget, he told himself, their lack of alacrity to follow him. He felt like a general without an army; his attention when he watched TV was not for his pleasure, therefore, but his martial art. He studied Ernie Kovacs like Moltke read Clausewitz, certain his hour was coming. The first TV show on which he would ever appear would be a call to the first battle, and he would fire a shot to ring around the global village.

THE FIRST SHOW happened to be *Night Beat,* Mike Wallace the host. Mailer got ready like a Depression fighter going into the main bout at the Garden on Friday night. Since his anticipation of such experiences had been formed by films of the Thirties, and his wife would in fact have their first baby in two weeks, since he also saw his opponent as tough, the pressure might be immense, but, as he saw it, he had to win. On just such trinities as poverty, babies, and tough opponents had John Garfield's movies been built.

Two decades later, Norman would not bother to think about a TV show until he was on it. He would learn over the years that the inner condition for projecting a firm and agreeable presence on TV was to be bored. Ideally, it was best to feel no more desire than a prostitute does toward her tenth client of the night. This abyss between the luminosity of the outer appearance and the void of the inner dark was a phenomenon unique to TV participants, and might be the very health of the malignancy; it was certainly what kept television such a formidable foe. How could one battle a process that accommodated the emptiest space in oneself?

Of course, on the afternoon before his first television appearance, Mailer was hardly in possession of such wisdom. Rather, he was as tight as a man going to the electric chair. As he tried to relax on his bed, adrenaline would take the best of him, and he would pace the room, and proceed to interrogate himself, asking the worst questions Mike Wallace might dare to ask, then search himself for the best of eight possible answers—a true fever! Young fighters dropped fights by squandering adrenaline before they reached the ring; they, too, were overcome with the monumental significance of whether they won or lost. Mailer, possessed of all the innocence of a protagonist who had never been on TV, did not understand he was entering no main event but was seeing the floor plans for a future mansion of Limbo; how was there a way to comprehend that he would not make history, but probably stifle it a little more?

It was a natural error to make. He could not, after all, get beyond the idea that he would be on the virgin air. If one said what one had to say with wit, conviction, and passion, if one said the unsayable, if people therefore heard something for the first time, passions ought to commence, buried sentiments should take life. But first he must be able to hold his own with Mike Wallace, and given Mailer's massive incapacity to stay cool, that was not automatic.

Wallace had become a success by way of a personality considerably at contrast to those of other TV hosts. His manner was not friendly but adversary. His straight black hair and craggy face gave off a presence as formidable as that of an Indian in a grey flannel suit. To this he added a humorless mind, of which he was utterly unashamed and therefore used as a weapon—no wit could slow his attack. His badly pitted skin suggested the unremitting stubbornness of a district attorney bitten hard enough in his youth to offer little mercy now. He was obviously not, like other interviewers, full of the very lard of companionability. Instead, Wallace gave the impression of wanting to interrogate further than the family manners of television permitted. Something of solemn, even heavy, purpose came off him, a distrust of human nature that was ready to focus on the probable insincerities of his guest. It was therefore a show that attracted jokes about the masochism of people who would accept invitations from Mike Wallace, but Mailer saw it as a rite of passage. If one wanted to make an impact on TV, *Night Beat* was the proper test.

By now, his own approach was as humorless as Wallace's. He had so fired up the household that his wife, in her ninth month, also took the show as a test and put on a low-cut black velvet dress, thereby mustering her best appearance. Her hair was as black as Mike Wallace's, and being a real Indian (Peruvian), she did her best to help her husband make a decent entrance. Indeed, she looked splendid and sent out the beauty of her pregnancy like a promulgation of status. Her mate might have been happier to come into a real ring with a trainer to massage the back of his neck, and he certainly had the ravenous flare in his throat that only true desire for whiskey can bring, but he also knew the happiness of the fighter—at long last, love! even if love is the pow of a glove and the rocketing of the ring lights. He felt equal to expectation itself—maybe he had never been so aware before of all of America out there—a pregnancy in the night itself equal to his awareness of a million homes tuning in to *Night Beat.* In later years, when everything was taped, he would not often feel the sensation that everyone was listening as the sounds came out of his throat.

Now the program began, that is, they took seats that had been set out facing each other, chatted a little—it was a false and most unnatural kind of chatting, like fighters meeting at a weigh-in—and then with thirty seconds to airtime, to Wallace's small but somewhat superior annoyance, Mailer poured his water back into the carafe, took a flask of gin out of his pocket, and decanted half of it into the empty glass.

He was sipping the gin as they went on the air. Since his mind had been overworked for hours, the liquor did not relax his tongue so much as disconcert his head and burn his stomach.

"*Life* magazine accuses your novel, *The Deer Park*," said Wallace in his opening and hortatory tones, "of dealing with immorality, alcoholism, perversion, and political terror in Hollywood. Why do you emphasize these themes?"

"*Life* magazine is a dirty magazine," Mailer blurted back.

The reply came out in a tense facsimile of the Texas accent he had learned in the Second World War. He knew that in his first swing at an American audience he had missed by a mile. Wallace's eyebrows raised like two eagles looking for prey. "Come, sir," said his expression, "that is not a reply."

Mailer knew it was not. He did not really think *Life* magazine was a dirty magazine. Rather, he thought it powerful, oppressively self-centered, and altogether ready to ignore his talents forever. So he throttled down and began a defense of the author's right to his subject matter, but his answer went on too long, and Wallace merely looked quizzical and superior in his lack of comprehension.

On it went, Mailer feeling uncomfortable through the next questions, too flippant for one, brusque on another. He was like a tight-muscled fighter and could not seem to fake Wallace the least bit out. His voice began to grow dogged. On the other hand, Wallace did not seem comfortable, either. Before one could expect it, the first commercial had come. One third of the show was over, and it wasn't going as Mailer had hoped.

So he raised the ante. If evil was the ability to discover what was good, and then proceed to destroy it, how many people were smart enough to be evil? Wickedness was more available. Wickedness consisted of raising the ante without knowing what the consequences might be. It was a way of asking for peril without a clue as to who would be hurt. Wickedness was therefore what he leaped into tonight. It was the best means of dealing with Wallace's imperturbable certainty that the people who ran things knew more than the people who didn't. Before long, Mailer was saying, "America is a great and prosperous country, but it is not a brave or noble country. Our leaders are drowning in conformity and act like *women.*"

(May it return a full sense of the Eisenhower period to recognize that our author was not near to suspecting a powerful movement would arise one day to banish men from public life for failing to refer to women as "persons.")

Wallace, being equally ignorant of the future, looked awfully confident. "Who in high office," he asked in his deepest voice, "would you say is so feminine?"

Mailer came back with the first happiness he was ever to know on television. "President Eisenhower is a bit of a woman," he said into the ring of night around them, and was certain he could feel the instant when the heart of a million TV sets missed a beat.

Wallace never looked more like an Indian. His eyes grew as flat as those of a movie Apache who has just taken a bullet to the stomach. "You don't mean it," he said. He, too, was listening to the silence in the living rooms.

They proceeded. The atmosphere was not dissimilar to the tension in a bar when a glass has been shattered and no apologies are made. They got through the half hour, and Mailer did a little better, and Wallace might have done a little less well, and the show came to its end. Mailer felt a great glow. He was immensely pleased with himself. In the elevator, going down to the street, his wife said, "Maybe we'll be dead tomorrow, but it was worth it." It is possible they never had a better moment together. They had been with each other for years, yet always fought in the animal rage of never comprehending each other. On this night, however, with the baby two weeks away, they were ready for one hour, at least, to die together. And, for once, Mailer felt like a hero. In those days saying something bad about Dwight D. Eisenhower was not a great deal less daredevil than deciding Jesus Christ had something wrong with Him.

Our new TV talent waited, therefore, on large and scandalized results from this Wallace show. Nothing seemed to happen. Occasionally he would run into an old friend who had seen it, who would now chuckle over his remark. Weeks later, he learned that James Hagerty, the press secretary to President Eisenhower, had asked for a transcript. That was all he ever heard. The phone did not ring with invitations from network executives to discuss a new show built around his controversial personality. Indeed, he was not invited onto other shows for a while. It was more than another year before he was asked by David Susskind to be on *Open End* with Dorothy Parker and Capote.

In Limbo, it was becoming less difficult to understand why he always felt an instinctive distrust of the memoir. The final insult to one's life was here reduced to viewing the moral lessons of the past through a film of slick. All the real, important, and (now that one was in Limbo) desperate meanings of one's acts slipped by, vague and smooth, almost without evaluation. The agreeable readability of the memoir (that property to make it perfect for magazines laid on airline passengers) was exactly what made it a torture now. Among the glossy walls of Limbo, moral facts were hard enough to hold without the frictionless surface of a memoir.

Ergo, the author found himself shuddering—even if he did not know why—at the recollection, let us say at the reconstitution, of his experience on *Open End*, retching indeed in that feverless ague where boredom is at last revealed as the ultimate protection against dread; yes, what better explanation for our family in the Utah desert staring at the video glass than precisely their awful, if unstated, decision, so gloomily made, that boredom was better than dread, that even an unpalatable mixture of boredom and dread was better than dread. Anything was better than that immense onsliding American dread.

For where but on this northern half of the continent,

in which the uprooted had looked for every frontier that could be broken (since people without roots discover peace in the sound made by tearing up the roots of others—how else account for the glee of airplane designers at the sound of jet engines sucking in the promise of the nearby air, or the joy of a young criminal on breaking and entering?), was there anything like it? Yes, if dread was the ongoing uneasiness of those whose roots were cut (which automatically made America the country most possessed of dread, since how many in this country could find the house where they were born?), yes, if dread was the communal American experience (how better account for the prodigious American fear of Communism in those years after the Second World War when the U.S.S.R. was at its weakest?), yes, if security against dread was the subtext behind every political movement in America—from right-wing fear of a conspiracy in high financial places to destroy Christianity forever, all the way over to the liberal terror that a violence was coming that would not be controlled by even the most sophisticated and enlightened social planning—then, given a dread so immense, it was easy to see why a man should be chosen for President so late as 1976 by way of a number of television debates that presented to the nation the first proposition of American politics: "When it comes to picking a President, Boredom is better than Dread!" So said the debates; and the power that was finally elected, the political movement whose real fortunes were advanced, was television itself, with its silent radiating argument that it was the best political program yet found for the tranquilization of dread: I will deaden you a little bit—you need it!

Contrast these grim conclusions, arrived at in the bowels, that is to say in the glassy coils of Limbo, and compare them with the innocence of Norman's early assumption that television was a platform from which to incite an ideological riot, a God-given opportunity (offered by the complacency of the establishment) to get the truth out; perhaps we can learn, after all, why he shuddered at the recollection of his appearance with Dorothy Parker and Truman Capote. He was now suffering from the derision of Limbo; few souls offend it more than the actor who believes he is becoming a protagonist, when he is only turning into a comedian. This unpleasant thought now as resistant to eviction as a tick on a hound's belly, he was able for a space to feel the grit of reality, and thereby found himself thinking again, and with considerable pain, of the expression on Dorothy Parker's face when he first saw her standing on the studio floor in Newark waiting for the arrival of Capote and himself. She had earlier been picked up in Manhattan, as they had been later, by a dark Cadillac limousine furnished by the studio to carry them to New Jersey for a few hours, and the expression in her eyes suggested it was outrage enough to be found in Newark!

Dorothy Parker was not to shine on this night. It was to be her first appearance on TV (and probably her last), and never may a video camera have been less disposed toward a talented writer. She was not, in person, a large woman, nor, by the measure of her literary wit, a bold woman. Anyone who had read Dorothy Parker and was expecting therefore a lady of immaculate wit and elegant savagery of tongue would feel on the consequence when meeting her a discernible shock. She was tiny, and she tended to be plump. She had bags under her eyes (from the fatigue of three good decades of drinking), and these bags touched the heart in a way that the wondrously lugubrious bags of few middle-aged women and only a very few middle-aged men can touch us. (Perhaps it is by way of the sorrows such people have bottomed out on while drinking.) Dorothy Parker looked like a bird at the mercy of every beast with teeth, a flurry of feathers up in a tree with wonderful bags under its eyes. And in the middle of her sad little face, sad as the ghost of tenderness itself, was her nose, the tip of which was always overpowdered. She would dust it once with a swipe to one nostril, and again with a swipe to the other, the overlay thereby suggesting the white painted button of a clown who is more lugubrious than his bulbous red-tipped colleagues.

She also wore a congeries of black shawls and garments that gave her the appearance of a British witch. The garments looked to have come from an attic. Since Norman had met her first in Los Angeles, seven or eight years before the evening in Newark, he could not imagine at the time where she found her clothes. Los Angeles was a city whose homes were conspicuously without attics.

She had also a manner that was gentle, full of praise, and treacherous as a scorpion. The day he was introduced to her, an agent who would later be powerful enough to deal with the memoirs of presidents and kings was in Dorothy Parker's room at the Chateau Marmont, and she gave the agent nothing but compliments, sweet compliments in a tender, threadbare little voice that spoke of his energy (boundless!), his goodwill (Christly), and his genius (large enough to sell one of her poor scripts; he, he alone, could do it!), and the agent having left in all the happiness of being given these compliments by a great American writer—he was still a young agent, pushy and prematurely bald—she said, as soon as the door was closed, "Agggrhh"—a clearing of the throat in pure facsimile of that first blast of the interior when vomiting—and added, "What an awful man. What a dreadful, dreadful man!"

After that meeting, Mailer always did his best to be the last to leave Dottie Parker's sitting room at the Chateau Marmont. Once, confessing to Lillian Hellman his concern about how Dorothy Parker might describe him in his absence, Miss Hellman laughed with the happiness she always felt in being generous, instructive, and superior to the innocence of her friends, and said, "Norman, everyone knows that Dottie will talk about you after you leave the room. It will amount to no more than the compliments she's just given you. Everybody who cares about Dottie knows that's the price to pay." And she laughed gaily.

Still, Norman wanted Dottie Parker's real affection. He wanted the lovely things she said to him on one side of the door to be repeated on the other. Since they got along very well, and he was always proud to invite her to his rented house, and pleased to be invited back to her poor suite at the Marmont (she had no money in those years—all her splendid short stories and no money!), he even entertained the illusion that she might offer a dispensation.

He would never find out, however. It was because of her dog. She had a boxer, not a year old, named Bruce, who had been badly mistreated by his first owners and now whimpered at harsh looks. Dottie did her best to give the dog courage—she raised him like the child she had never had, ready to take Bruce up each rung of the mile-high ladder that climbs from abysmal cowardice to transcendental bravery; she would tell her friends of each new advance Bruce and she had made. One would learn that Bruce had gotten up his nerve sufficiently one day to take a walk around the block with her.

These methods seemed intolerably slow to Norman. He kept telling her that she ought to bring Bruce over to meet his dog. Since Dorothy Parker had by now made friends with Karl, she was tempted. Timid people were invariably proud of such a friendship, for Karl was a black German shepherd, huge and ferocious to strangers. Since his first allegiance was to the house, his master became whoever lived in the house (and, incidentally, fed him). Norman, renting the place, was therefore the master and thus became the only man in Laurel Canyon who could drive up to the house, get out of his car, and not be attacked. No one else dared. Our best-selling young author used to run out to the driveway when guests came; otherwise, visitors were obliged to cower inside their vehicles while the dog lunged at the windows. Once Norman got out, however, and barked "Karl!" the shepherd immediately subsided—he could now recognize that these intruders would be acceptable to the house. After they were admitted to the living room, Karl would even go over, sit down on the floor, and make a point of shaking hands with them. Delighted that the dog had become well-disposed, the new arrivals would make much of this. Karl would extend his paw for a second greeting. His tongue would loll, and he would pant a little. After the third handshake, Karl would get a hard-on. *"Karl!"* his master would say, and the hard-on would go away. Under such circumstances, it became easy to think of Karl as an overrated danger.

So when Dorothy Parker finally decided that she might bring Bruce to the house, Mailer thought it was a terrific idea. He would take care of bringing them together peacefully, he promised. Good to his word, he put Karl away in the bedroom before Dorothy Parker arrived with Bruce. The boxer, however, did not want to quit the car. He could hear the other beast.

"It'll be all right," said Mailer cheerfully, "once Bruce is inside. Karl is really lonesome."

Coaxed within the house, however, Bruce urinated on the carpet. It was not difficult to forgive him. In the bedroom, Karl was making the sound a cat will make when a dog is near and she is ready to fight for her life. It is a more awesome sound when it comes from a large hound.

"Don't you think this is a mistake?" said Dottie Parker. "I can put Bruce in the car and go home now."

"That would be a loss for Bruce," Mailer answered. "Why don't I put Karl on a leash, and bring them together slowly? It'll be so good for Bruce if he and Karl can make friends."

This liberal and romantic expectation prevailed. Karl was brought out of the bedroom. The dog came quietly on his tether, and looked at Bruce without a sound. He sniffed the other's nose. A pause. Bruce peed. Karl's head went off in a roar as loud as a howitzer. His mouth flayed the air with strings of saliva. At this point his master was sufficiently outraged to yank him back by the neck, and manhandle him to the bedroom. In the space this took, Dottie Parker was out the door and into her car. Before she could drive away, Norman came running out to find Bruce collapsed on the rear seat. The dog's limbs were vibrating. It was that quiver which will not cease until the nerve dies.

"I said it wouldn't work," said Dottie Parker face-to-face. She imparted this in a tone of whole reproach, as if to tell Mailer that the worst damage a friend can wreak is to convince you to be brave when you know you are right to be cautious, and then proceed to prove that indeed you were right. It throws you back even further upon yourself when that is no place to be.

As she drove off, he knew that he had wanted to bring Karl from the bedroom because he wished to show how well he could control the dog. Besides, he had been curious of the outcome. He noticed that he did not feel very sorry. Only chagrined. But, then, the wicked are famous for offering small remorse.

HE WAS NOT to see Dottie Parker again until the evening they would meet on *Open End* eight years later, and when he arrived at the studio in Newark and saw her waiting—older, more fragile, close to desperately nervous, and certainly not cordial—his feelings were only in a small way hurt that she was not happy to see him again and showed an obvious preference for Truman Capote. He was, by his own measure, not the same writer she had known before. He had met her, after all, in the season after *The Naked and the Dead,* when he had gone out to Hollywood, and much had been made of him. Now, after that early success, he had endured a number of years in which people spoke of him as a failure. Since he had written two novels, *Barbary Shore* and *The Deer Park,* that he thought had received unfair reviews, he had a view of himself that was at odds with that held by others, and he did not like to be snubbed by people who had been friendly to him ten years before. All the same, it gave justification to his anger. So her unfriendliness left him in a well-balanced position. If it came to it, he would not have to be too friendly on the show.

David Susskind also gave small welcome. "We met a long time ago," Susskind said.

"I remember."

"I do, too," said Susskind in a chilly voice. It had been at a party on a very hot summer night in 1948, and they had talked on a Manhattan roof. Susskind was then a young agent, and he had wanted to sell *The Naked and the Dead* to the movies. Mailer had made fun of him. "Don't you understand?" he had said to Susskind. "*They* can't make it into a film, and I don't want it made."

Of course, years later, Mailer sold it, after all, and *they* had made a very bad film. So he felt at a disadvantage with Susskind now, and said, "Maybe I should have listened to you."

Susskind's face showed what he thought of people who were rude years ago and ready to flatter now. Their relations for this night were certainly not off to a good start.

He had only Capote. They had gotten along well on the drive out to Newark, each—they had not met before—intensely curious of the other. Since Mailer's wife accompanied him again, and again was looking her best, he had made a point of having her sit between Truman and himself. Since she was twice Truman's size and had the overpowering sexuality on occasion—this was such an occasion—of her burgeoning Latin blood, it had been an advantage equal to playing on one's home court.

Truman did much complaining on the trip out. "I didn't want to do this show," he said in a dry little voice that seemed to issue from an unmoistened reed in his nostril. "I told Bennett Cerf it was a mistake, but Bennett thinks television is going to be very important for selling our books. I hope he's wrong," said Truman Capote, and laughed.

Once the show began, Norman thought himself splendid. Unlike the night with Mike Wallace, he felt at his best: thoughts came clearly, he was full of energy, and the others seemed bewildered. Susskind labored in the early minutes to put Dorothy Parker at ease, but he was not able to succeed; she simply would not trust him. She spoke in a pained and quavering voice, hardly able to make herself heard. Truman did not add much, and Mailer, pleased with his powers, began to take over.

Soon he and Susskind were doing most of the talking, that is, most of the debating, for it was obvious Susskind was irritated by what he had to say. Whenever Norman would launch on a flight of what he considered well-stated criticism of society, Susskind would make a point of looking at his watch. Whenever he could, Susskind tried to give time to the others, but it did not work. The others seemed apathetic. Once, while expatiating on his aversion to politicians, Norman Mailer flew so high as to say, "They're all whores."

Dottie Parker interrupted then, to say with a little real force, "That's a sweeping remark."

"Well," said Mailer, "it may be sweeping, but it certainly is true."

It put Dorothy Parker back into silence—who was she, after all, to defend politicians?—and it annoyed Susskind more. After close to an hour of TV time that felt wholly enjoyable to Mailer, he began to grow irritated at Susskind's efforts to slow him up. Since his host seemed determined to shut off this most interesting part of the show, Norman could see no reason not to allow the host to hang himself, and so he ceased speaking. Susskind began immediately to draw upon Truman Capote, and Capote, having digested a few of the peculiar processes of this odd medium and measured the possible fit for himself, began at last to speak, and was, Mailer thought, not unamusing. He laughed encouragingly at Truman's remarks; he offered attention. Truman was so tiny that something gallant came to you from the fact of his existence itself. Mailer felt generous indeed. Few moods are as charitable as this sensation of being physically superior to everyone in the room.

In such a state, feeling handsome, vital, and more interesting than anyone had a right to be, he got into a discussion with Truman about the merits of Jack Kerouac. Since Mailer was not without his jealousy of the large attention paid Kerouac that year, he gave a defense of *On the Road* that was built on the basis of calling Kerouac "Jack"—that is, he was two thirds for Jack's virtues and one third against Jack's vices.

Capote detested Kerouac. As Mailer grew benign, Capote grew precise. He rose at last to his own peroration and invoked the difficulties of the literary craft in contrast with Mr. Kerouac's undisciplined methods of work. Finally, in a tone of fearless and absolute severity, Capote said: "It is not writing. It is only typing."

"I agree," said Dorothy Parker in a hoarse voice.

"Well, I don't," said Mailer, only to give a limp defense. He was empty of vast indignation at this dreadful put-down of Kerouac. He even decided it might look good to let Truman have his winning little moment; certainly it would look better than trying to hog the show.

Open End came to a finish, and they walked off the set. A few technicians were studying the kinescope of what they had just done. Therefore, Parker, Capote, Susskind, Mailer, and Mrs. Mailer stopped to watch a minute of it.

Dorothy Parker did not look good on TV. She took one peek at herself, gasped, and said to Susskind, "No, I really don't want to see another instant of it." Solicitous as Saint Peter, Susskind led her away.

In contrast, Norman Mailer had the pleasure of seeing himself as he wanted to look. The kinescope caught him as he was making a point, and his face looked forceful, his language was good; yes, he appeared even better on TV, he thought, than in the mirror. "Didn't I tell you," said Truman, "that you're terrific on this? You're *telegenic*!" It was true. But then, during every intermission, Truman had been telling him how splendidly he was doing.

Truman groaned when the camera shifted to him. There was a medium shot of Truman looking a hint bewildered at the beginning of the show, and he winced as he heard his voice. "I told my publishers I shouldn't go on," he declared with annoyance.

In the limousine on the way back to Manhattan, Truman kept saying, "I shouldn't have done it. I didn't want to, I've never appeared on television, not once—even though I've had many invitations—and I certainly oughtn't to have been there tonight. What I can do is special, it's *very* literary, and I shouldn't attempt to *intrude* my personality. I'm not good at that like you, Norman. Television is good for you."

Mailer lay back in the limousine, enjoying a winner's ride, and his wife said encouragingly to Capote, "You were good, too, Truman. You really got better as it went on."

"Do you think so?" asked Truman. "Do you really think so?"

Now both Mr. and Mrs. Mailer assured Truman Capote that he was better than he thought he had been.

"Well, at least," said Truman, "I was better than poor Dorothy Parker. When she looked at herself afterward"—he shuddered in sympathy—"wasn't that *dis-*

asterville?" They laughed, but not in pleasure at her woe so much as in wonder at the clear lines of judgment itself.

Truman could not quite get out of his bad mood, however. Before they said good-night, he suggested they have a drink with him at El Morocco. Before long, they thought to have supper. Despite all the sandwiches served on *Open End,* they were starved. Truman insisted on treating. A prince, he seemed to hint, must play the host after a sorry loss—it is the only way to come close to one's blood again.

It was late on Sunday night, and they had the place to themselves. Under the solicitousness of the waiters and the captain, and finally the manager himself—who all came over to assure Capote of their unremitting attention—under the spell of this wholly superior service Truman was obviously accustomed to receiving whenever he went to El Morocco, his spirits not only revived but turned charming, and he began to offer an attention to Adele Mailer that had her believing in the unique properties of her attractiveness and her wit; yes, Truman was charming. Mailer saw why he had become the in-house author of the most important hostesses in New York, and envied him honestly, the way one good athlete will have respect for another. Just as a pitcher will think, "If I had as much stuff to my curve, I'd win more games because of my superior ball sense," so did Mailer think wistfully that if he could get to a few of the select parties Truman got to, why, boy, he'd have more to say about society. Of course, how could you be a radical yet an intimate of the top drawer of the world?—that was the iron warp of irony itself.

They had a fine time. They discovered unexpected points of agreement, and lively places to disagree; they had each had such curiosity about the other after all these years—their first novels had come out within a year of each other's; both had been celebrated almost instantly. It was a grand evening, and at the end, Truman announced they would be great friends forever, and thanked them for taking him through what would otherwise have been a terribly depressing night—"God, was I awful on that show," he said as they parted, and he made a face to simulate the gargoyle of godawfulness.

There were not many nights when Norman Mailer went to bed so pleased with his wife, himself, and what they were able to do for each other as on this night. He woke up early in the morning, enjoyed his breakfast, and could hardly wait for the phone to ring and the praises of his friends to begin. But when he had heard from his parents (who loved the show), from an aunt and uncle (loved the show), from his sister (who liked what he had done—"generally"—and was somewhat taken with Truman—"What an interesting personality he has, really," she said), and when he had gotten tired of waiting for friends to ring, and called up one instead, there was a pause at the other end, and then the friend said in sorrow, "Oh, man, did Truman take you!"

He called other friends. They seemed not to have seen the show, and this left a pall equal to the message of the medium itself. He went out on the street for a walk; since he was living in Greenwich Village, it was not hard to run into a few acquaintances. One stopped, saluted, and said: "It is not writing. It is *only* typing," and gave a long phlegmy laugh as he went on down the street. Another said, "Truman—too much!" A third said, "Could you get me to meet Capote?"

More serious friends were balanced. "In terms of polemical points, I think you scored the most. Of course, Capote does suggest real authority."

He called Truman a couple of days later.

"Yes," said Capote, "isn't it the *strangest* thing. Everybody has been telling me how marvelous I was. I can't believe it. 'Norman Mailer is the one who was marvelous,' I say to them, and they reply, 'Truman, you were *wunderbar*. Don't sell yourself short.' 'Honey,' I told this friend, 'I been short all my life.' " Truman guffawed with pleasure even more profound than on the night at El Morocco. "It certainly is a mystery," he said complacently.

Mailer couldn't bear it. He finally called *Open End* and asked to see a kinescope. It became his introduction to our hypothesis that television is not a technological process that reproduces images of real life by way of electronics, but is, rather, a machine (more or less cosmically operated) to anticipate the judgment and/or anathema of Limbo; technicians collaborate by the use of camera angles.

He discovered that the nice close-up he had seen of himself immediately after the show was one of the few close-ups they had chosen to give him. The more he talked, and the more Susskind looked at his watch, why, the more they relegated Mailer to medium shots and long shots. There is something pathetic about a man speaking at the bottom of a long shot: he does not sound convinced by himself. He is, after all, at the end of the tunnel.

In contrast, they gave many close-ups to Truman. Capote did not look small on the show but large! His face, in fact, was extraordinary, that young-old face, still pretty but with such promise of oncoming ugliness; that voice, so full of snide rustlings and unforgiving nasalities; it was a voice to knock New York on its ear. The voice had survived; it spoke of horrors seen and passed over; it told of judgments that would be merciless.

Watching the kinescope, Mailer realized at last what an impact Capote had made on the television viewers of New York. They had never seen anyone remotely like him. Once Truman finally opened his mouth, the camera never left his face. The camera would stay on Capote even as Norman was speaking.

His own arguments, which had sounded so forceful to him a few nights before, now seemed vague and, with his beard, pious. (He shaved off the beard a week or two later.) Talking about large matters on television with all the passion and all the lack of specific detail these immense theses encouraged, speaking in a voice that came out of the depths of a medium shot, proved to be not nearly so exhilarating to a viewer as it had felt to him in the middle of articulating his thoughts. His obvious pride in his ideas now seemed fatuous. His physical superiority was gone. He was only intense, vague, and a bit out of focus. Whereas Capote took his unforgettable personality and added to it practicality, common sense, and pride. In so misty a medium, the best gift a guest could provide was pearls. The certainty of a pearl! Yes, to light upon a problem and come up with a good one-line answer was to produce a pearl of the mind. How much interest could viewers have in a point that took five minutes to be made? That was something they could not do. It spoke of too good a college education. So they preferred people with one-line answers, barroom knockouts, one-punch pearls. There is no man or woman on the street whose mind will not produce a pearl from time to time. They loved Truman because he had given them "It is not writing. It is only typing."

From now on, thought Mailer, he would not try to show how intelligent he was; he would look for pearls. That became his latest hypothesis: The light from a TV screen was flattering to a pearl.

Of course, he was always miserable without a hypothesis. The hypothesis could later prove invalid, but it had to be interesting enough so that he could live with it for a while. That the orgy would benefit humankind more than the family was one of his early hypotheses, Bolshevist as hell. That the war in Vietnam was a bad war because rich boys should not fight poor boys unless there is equality in the weapons was another, which offered the best of conservative principles: The rich had better be terrific. Whether the particular hypothesis was good or bad, however, he leaned on it. There was something so rigid in his upbringing that for reference he had great need of a straight line—a straight line is, after all, the very first of the hypotheses, and, according to Einstein, wholly incorrect, since no line can run straight in space, given the curly route of a light wave and the oval field of magnetic force. Mailer sometimes wondered if he was one of the first to recognize that 186,000 miles a second was not the true speed of light, but only its calculated progress along a nonexistent straight line. If you measured the speed of light by the time it took to wind through the waves, then a beam might travel up to a million miles a second in order to traverse 186,000 miles. That, doubtless, was one more reason why the sum of a million was magical to Americans.

Such were a few of Mailer's conclusions after his defeat by Capote.

The next time they met, Truman had a new assurance to put on top of the old one. In these few weeks, he confessed, there had been numerous invitations, all turned down, to appear on other shows; a difference in Truman's idea of himself had begun to emerge. That personality he had presented, with all his early bravado, to a most special part of the world was, it seemed, going to be accepted by all the world. He obviously felt stronger already. And Mailer, stunned as any confident contender who has been abruptly knocked out, now felt, measure for measure, weaker. It was obvious to him (for no one understands the future reactions of a snob so well as an unsuccessful snob) that never again would Truman spend a night trying to recoup the losses of his ego out of *this* Mr. and Mrs. Mailer.

YEARS LATER, Norman would decide that television was one of those hostesses who are most impressed if you are somewhat unimpressed to find yourself in their homes, and that was easy enough to handle. It was just that on television your real hostess was the spirit of the electrons. While nobody was in command of enough physics to explore the idea that some men and women tend to attract electrons while others most certainly repel them, it seemed reasonable to suppose that electrons might well conduct relations with more than a few forces, satanic and divine, and so could be working in unseen harness to focus or ruin your TV appearance. Translated, that should state: The telegenic have bioelectric relations with the tube, and these relations may yet be measured. Of course, Norman, biased to the depths of his unexpected defeats, was convinced that zombies were the first to benefit from such relations.

These were conclusions he came to with reluctance. He was plagued in those years by the divide between his absolute, if somewhat improbable, sense of intellectual mission and his incapacity to work at anything like the necessary pace to accomplish his wonders. He thought, without admitting it to himself, that television might be a way to express ideas that would otherwise never reach an audience, not if they had to wait for the writing. (Obviously, he wished to lift himself to heaven by the power of his speech—a misapprehension of cosmic reality he only came to recognize when his later writing, so highly praised, raised him no higher than Limbo.)

Of course, he had an idea not easy to express under any quick circumstances, and hardly by way of the pearl: he saw a conspiracy in society to destroy existence. Being blessed with no ear for the finer sounds of nature, he did not hear the drip of pollution in the conduits; he only felt the stifling of his own existence. That, of course, is enough. It may be even a sign of health to sense danger to nature in every little death of oneself.

So he wanted to be on TV; so he told himself he would be on it to shout the alarm. So he saw every appearance where he did well as a decent effort committed to his side of the gods; yes, he would do his best to get the great dialectic of society unfrozen once more—that was the dialectic where the people knew enough to affect even the most skillful manipulations of the establishment—he would, yes, try to get such a dialectic going. A mature democratic society was worth zero without it. He did not understand that *something* was calculating how to cancel more than one dialectic.

Still, these were the early years of TV. He had a few moments when he felt as if he had broken through. It was not entirely a string of disasters. He had, after all, the illimitable freedom (in early 1960) of saying on Irv Kupcinet's show in Chicago that he thought J. Edgar Hoover had done more harm to America than had Joseph Stalin, and so had the pleasure of being the center of attention one early morning in Chicago; Kup's show was big news to the local audience, for it gathered all the visiting firemen. An amazing number of people would come home on Saturday night and go into four a.m. of Sunday morning watching *At Random.*

"Yes," Norman repeated, "J. Edgar Hoover has paralyzed the imagination of this country in a way Joseph Stalin never could," and the other guests put their hands in the air to state that they were not opposed to J. Edgar

Hoover. Phil Klutznick, the former president of the B'nai B'rith, was a guest, and Robert Briscoe, the lord mayor of Dublin, was another; a chief from the national police of Sudan made a fourth; Elaine Shepherd, a reporter who had traveled with Eisenhower in South America, was the fifth; they all were in a hurry to swear their allegiance to Hoover.

Later, after the show, he heard that a large number of calls had come in, forty-eight in fact. He had always assumed that thousands of calls came to such shows every night. He did not yet know that television did not send you to the telephone, let alone the barricades. Still, forty-eight calls had come at two in the morning. Twenty-eight opposed what he had said; twenty approved. Since no call was counted unless you left your name, it meant that twenty people (with whatever number of false names among them) had been ready to agree. That count was as agreeably astonishing to Norman Mailer on that night as Gene McCarthy's forty-two percent against Lyndon Johnson's forty-eight percent in New Hampshire would seem apocalyptic to the media in '68. Mailer felt an intimation for the first time that the Sixties might bring a few changes to the old game.

Eighteen years later, obtaining his F.B.I. file by way of the Freedom of Information Act, he would discover that out of three hundred photocopied and highly deleted pages sent to him, more than thirty were devoted to that single appearance on Kupcinet's show.

TO: DIRECTOR, F.B.I.

FROM: S.A.C., CHICAGO

SUBJECT: NORMAN MAILER, PANELIST
IRVING KUPCINET
"AT RANDOM" TV PROGRAM
MARCH 20, 1960
INFORMATION CONCERNING

NORMAN MAILER is an admitted "leftist." He is also an author, his most popular work being the novel "The Naked and the Dead." His remarks clearly reflect his animosity toward the Bureau.

The enclosed text clearly reflects that other noted panelists on the program definitely and specifically disagreed with and refuted his observations.

A number of persons in Chicago have personally commented to me that Mailer made an ass of himself on the program, and all such comments received have been definitely pro Bureau and anti Mailer.

Re: Norman Mailer

During the program, the participants engaged in a discussion relating to the Federal Bureau of Investigation. These remarks are being included below in summary form and at pertinent places verbatim.

Mailer talked of the United States as the most powerful nation in the world after World War II with the exception of Russia, and both these countries, who had no tradition of ruling, fought about who would rule. The United States began to ape the methods of Russia with the tactics of the Communists being taken by Madison Avenue, the term used to indicate the advertising agencies who run the mass media. Mailer went on to say:

I said the other day in an interview that we have a very subtle totalitarianism here, a pleasant totalitarianism, in that we don't have concentration camps. We don't have a secret police everywhere. We do have an F.B.I., which is one of the two religions left in America. There's the F.B.I. and medical science.

Elaine Shepherd: And hooray for the Federal Bureau of Investigation.

Mailer: Well, I would say three cheers against them. I am for putting them down. I say let's have no more of the F.B.I.; they are running the country.

Shepherd: You can't be serious.

Mailer said he had only one more comment to make with regard to Miss Shepherd's saying the Communist aim is to make America soft.

Mailer: This country was softened up, but not by the Communists. It was softened up by the F.B.I., by Joe McCarthy, by the House UnAmerican Affairs [*sic*] Committee, by the peculiar, subtle psychological reign of terror that went on in this country five years ago. . . .

Klutznick: People have abused J. Edgar Hoover, and there are times when I have been angry at him myself for some things he said, but let's put it this way: he has given us a kind of police administration that people like Inspector Hamid and others all over the world respect because of its efficiency and, more than its efficiency, its peculiar devotion to staying within that law which it is permitted to stay within and nothing more.

Shepherd: At really comparatively low salaries. These are dedicated men.

Klutznick: This is very important to remember.

Mailer: This is marvelous, but I would like to say something. You speak of efficiency. You speak of the low salary the people have who work for it. . . . I will also say that the concentration camps at Buchenwald were run in exactly the same way; they were terribly efficient.

Klutznick: This is not true.

Mailer: They were efficient. The people who worked there received low wages, relatively, and they probably kept within the directives that were issued to them.

Shepherd: Now, that is beneath the dignity of sophistry.

There was a momentary digression with some remarks on the meaning of sophistry.

Klutznick: No, but look. Before you get—when an American says that the F.B.I. operates like those forces did who ran Buchenwald—I read the records of Buchenwald—

Mailer: Now I didn't say—I didn't say—
Briscoe: You said they were efficient and with low salary.
Mailer: I didn't say that the F.B.I. is a Buchenwald. I only said that the virtues you ascribe to the F.B.I. can be ascribed to the concentration camps.

The discussion turned to the proposition of the police as a necessary evil.

Mailer: . . . the man in the street does not have a specific knowledge of the right things and the wrong things that the F.B.I. does. He has a general knowledge of them and a general fear of them and a general respect for them which, I say, is too great. Because what I am getting at is this: there has been so much damn fear in this country for the last five or ten years . . . that people in this country are being worn down, and they are being softened, and they are turning apathetic, and there's getting to be very little passion left in American life, and this was about the most passionate country in the history of the earth at one time, and the most adventurous and the bravest and the boldest, and it has become a cowardly, sleek, soft, fat country.

Klutznick: I would agree with some of that.

Mailer: What I am saying is that this is exactly the climate which the Russians adore, and if they ever come over here, the first thing they would do is that they would have their secret-police chief get together with the F.B.I., and you would be amazed at how little time it would take—if the same people who are working for the F.B.I. now—some of them couldn't do it. Others would make the transfer over, and they would still be working for the F.B.I. except it would be a Russian F.B.I. We are exactly creating a totalitarian climate in this country which will make it possible for the Russians to come over and take us with no pain at all.

Shepherd: You are questioning the patriotic integrity of our federal agents?

Mailer: That they are patriotic I would not deny for a moment. I am sure they are all passionately in love with America, and maybe a little bit too much in love with America. The thing is that I distrust it. You know if you met a man who is a very big

man who is obviously very powerful and very good-looking, and he starts to tell you how he is a great lover and a great fighter, and so on, we all begin to have a little doubt about the man. You know he doesn't have to talk about it if he's really got it . . .
Klutznick: This is a very different subject, and you and I will agree on a lot of this.
Briscoe: It is a different subject.
Mailer: I think it's all part of the same subject.
Kupcinet: I can agree with Norm that we should stand back and take a look from a distance; don't be swelled up by anything you read about anything. Ask questions about everything, and this would be a better democracy.

In another part of his file, Mailer found a letter from S.A.C., New York, to the director of the F.B.I. It was dated March 25, 1960:

> Information has been received that Mailer, in an appearance on the Chicago television program "At Random" on 3/20/60, advocated the abolition of the F.B.I. He said the F.B.I. was too efficient against Communism and was connected with McCarthyism and the House Committee on UnAmerican Activities. He also said that the F.B.I.'s efficiency "created a psychological atmosphere on the part of the public and that the public is being protected from Communism" and that the public, and not the F.B.I., should be allowed to decide for itself what is good for it.
>
> You are requested to bring the investigation of Mailer up to date.

A sentence about two lines long followed this remark, but it had been deleted before the file was sent to Mailer.

HE COULD HARDLY have known the F.B.I. was taking such interest in his few remarks—he did not really believe in his own paranoia—yet somewhere in those marketplaces of sleep where the knowledge of one dreamer passes over to another, in those psychic flights that lift from the mind and are on the air, maybe we reach another human despite all the crackling and confusion and outright din of all those amplified waves electronically induced by torture of the air itself; maybe our own thoughts, often so noble as to originate out of the magnetism of human mood itself, also pass back and forth like radio waves and enter our sleep despite the interference of every electronically communicating machine, yes, come to us through all the jammed air and defoliated fields of lost prayer. (Is that any more difficult to believe than that one corporate machine can pass a message to another when they do not touch?) So some obscure but real message may have come to him that the F.B.I. had bit on his words and he had pricked their blood; afterward he was fortified in the notion that TV could be a forum for ideas, at least for ideas that came in the shape of pearls—J. Edgar Hoover has done more harm to the minds of Americans than has Joseph Stalin—maybe that you could change people's lives and give a shift to society: you need only be good enough to do it.

So he kept the idea that he would yet learn to handle himself on the most obdurate and repellent medium in which he ever worked. He certainly kept appearing on TV shows. As the years went by and a little more literary work began to get done, there seemed always a new book for which it was natural to go on TV. Of course, TV did not help his books to sell—not with his personality!—but still he went on. He learned that an appearance on TV left you feeling as agreeable as a session with a doctor and an X-ray machine; he learned that the taste of aspirin (was it ashes?) would follow in the mouth; he learned that you formed no friendships with other guests on a show, for everyone split the moment a program was done, as if they had all been participants in a bloodless, sexless, and faintly disgusting orgy—some unheard whinny of sanitized shit was in the air.

He learned that the pioneer days of TV were indeed over, for only one experience on TV had been great, and that was to be on *live,* on to feel the unfelt oncoming reaction of millions of people all at once—that was gone. Now every show was taped, which is to say ninety-nine out of a hundred shows were taped, and the others were either early in the morning or news shows, and the newsmen read their texts. It meant that you were usually recorded at ten in the morning or six in the evening for a show that might go on at midnight that night, or on a night the next week, or in half a year, and so you were skewed out of time. The subtle emotions of a previous day and another hour would go into the midnight perceptions of an audience, speaking to them at that moment even when they knew better—that you were speaking to them at that moment even though you were not, and that was a subtle poison to slip into the perceptions of anyone studying a TV set, not to speak of the chop in the waves on wherever was the sea of time.

Yet if nearly everything but the early-morning shows was now taped, and even the *Today* show had a seven-second lag between what you said and the instant it reached the audience's ear, time precisely for "Fuck the President" to come out as "Bleep the President"—

sounds, in any case, that only a madman would be able to make at seven-twenty-two a.m.—if you knew as you got on the air, therefore, that the air was not the air, any more than the last shiver of a transcendental act is going to pass through the lining of a condom, and that some technician had his fingernails on a knob between you and the air, then you were not live, you were captive. Words good enough or conceivably great enough to move the doors of history in any future time of stress would never get into the transmitter. If a subtle totalitarianism or an outright fascism came to America, then the instrument to control TV had already been installed—your words were taped, and on occasion they were bleeped. Even a hero looked like a chicken when his words were bleeped.

Still, he went on. He could not rid himself of the notion that it was a technique worth learning. Besides, it is pretentious to look for virginity in the midst of the promiscuous. So he went on Joe Franklin, and on Mike Douglas and David Frost and Cavett and Merv Griffin and Carson, and even appeared on Steve Allen in a ghost of the old show. He went on CBS and NBC and ABC, he went early in the morning to be taped live—that is, with a seven-second hitch—and went in the late afternoon to be seen in the middle of the night, and he went on channels 5 and 9 and 11 and 13, and station KNET in Los Angeles, some such letters, and other letters for Detroit and Houston and San Francisco and Cincinnati, New Orleans and Chicago, Denver and Atlanta and Toronto and Washington and Philly; he went on with athletes and singers and dress designers and comedians, with ingénues and the inventors of novelty products, with doctors who knew diets and doctors who could speak of sex, he went on with worthy and self-serving politicians and lawyers for newsworthy causes, and once he even went on with Mother Teresa and Malcolm Muggeridge; he took what was offered, seeking to refine a technique whose first lesson he might never have learned, yet something in his stubbornness insisted on keeping the idea: he must use every means to advance his idea, and rarely did. And over the years, his idea changed. For if in the beginning he had seen a conspiracy by society to stifle existence, he came in time to think there was a conspiracy in one part of the cosmos to destroy the existence of the rest of the cosmos, and how do you get such an idea across when it involves a train trip to Philadelphia, a half hour with Mike Douglas and Rocky Graziano, and back again? He might have been a wrestler, working one night in Atlantic City, the next in Pittsburgh; he took falls that would have broken his back if he had had a literary sensibility left; he listened to Joe Garagiola once on the *Today* show reading from *Why Are We in Vietnam?* "Bleep," said Garagiola, pretending to recite from the book, "I want to bleeping tell you that that's bleeping bleep when the bleeps hit the bleep." It was proper payment for going on the *Today* show to try to advance the fortunes of *Why Are We in Vietnam?* On the other hand, it was just a wrestler's loss; he was supposed to lose that event.

He went on wherever they invited him. He would go on Bob Cromie's *Book Beat,* and act like a gentleman; he would go on Susskind again, and again; he was on with Buckley once in '64, and Buckley finally allowed that Russell Kirk was right in calling Norman Mailer a freak, and Mailer, in simple pain (you weren't supposed to get hurt in wrestling), said, "If you think I'm a freak, why did you invite me to dinner?" and Buckley shrugged and said, "No matter," and Mailer never forgave himself for not throwing a punch at that moment (although he wanted to get over the reputation he had for throwing punches at people). He waited for the next show they were on together—it was Les Crane who had the new interview show that year—and Mailer called Buckley a very bad word, and Buckley, as if he had been waiting a year for this, calmly raised his hand to Les Crane and said, "I assume you're cutting this out of the show," and Crane said of course he would, and a few hours later, since Buckley was drinking with Mailer in the best goodwill, Norman apologized, for he knew the word he had called Buckley was not true, or in any case—how was one to know of anyone?—not fair, and finally probably not a sin, and Buckley accepted in good grace. Hours later, Mailer began to brood over the little fact that Buckley had not apologized in turn, and then he began to wonder if one of the reasons for the Buckley family fortunes was that they did not go around exchanging apologies for too little. Buckley invited him once onto *Firing Line,* and he had the pleasure of telling Bill that he thought Fidel Castro and Charles de Gaulle were the only heroes he could find in the modern world, and Buckley may even have missed a breath over that remark. Mailer was beginning to take pleasure in his pearls. Like an old pederast who keeps warm with the memories of a few divine young phalluses he has seen, Mailer lived with the memory of a few of his remarks. Once, in London, on David Frost's *Frost on Saturday* show, indeed on the night he met Frost, David had come at him like a British heavyweight mauler—no great skill, but plenty of shove: "All right, Norman Mailer, you've written that sex can be killing. What do you mean by that?"

"David, I don't think one should talk about sex on TV."

"Why don't you?"

"It's like talking about your wife."

"Now, Norman, you said sex is killing, you wrote that right here, and we want to know: what is so killing about sex?"

It was a live show. They still had live shows in London in the late Sixties. He could feel half of Britain gathered around the hearth; the other half were right next to their TV sets. He was getting ready to speak to this other half.

"Tell us, Norman. Don't be coy. What is punishing about sex?"

"David, do you want me to unzip and lay it on the table and show you the welts and the warts?"

It was his best shot to a TV interviewer's eyes since he dropped Eisenhower-is-a-woman on Mike Wallace. Frost looked like a British heavyweight who has just been tagged but somehow does not go down. His eyes blinked. The two men sat there looking at each other over the sputtering of the British hearth.

Frost went on to do the rest of his show and do it well,

engaging in a dozen arguments with a panel of management and labor-union men: he took a good punch. And Johnny Carson gave a good punch (in the clinches), as Mailer would also discover on the occasion in '72 when he announced he was for McGovern while sitting next to Carson. During the break, as the canned commercial went on, Carson leaned in and said, "You really like McGovern?"

"Yeah," answered Mailer.

"Nah," said Carson. He leaned in at the advanced angle of a slim man about to slip you a tip on a horse: "I was talking to six friends, and on an impulse, I don't know why, I stopped and said to them, 'Who does McGovern remind you of?' What do you think all six said?"

"What?" asked Mailer.

"Liberace," said Carson and leaned away. They were on the air again. "Norman," said Johnny, "you were about to give us your views on McGovern."

Mailer did his best, but Carson had taken two percent forever from his genuine respect for George McGovern.

Yes, out of the sum of spiritual wrestling matches and invisible boxing fests, out of that accumulation of hours over the years that felt like exposure to the subtlest cancer of them all (and had they ever done a correlation between the purchase of TV sets and the rise in cancer? —no, *they* hadn't; no, *they* wouldn't!), he came, more and more (to his own embarrassment), to treasure his TV pearls. They took some of the taste of the aspirin away; they had to; nothing was so depressing as those predictable draws TV encouraged; you were polite, the host was polite, an idea went back and forth and died.

All the while, out in the vast pall, *something* was coming out of the sets. Anything to prevent that.

One time on Merv Griffin, back in 1967, he attacked the war in Vietnam as an evil war that would bring every horror in this country up from the sewer, and went on to describe America as a huge drunk covered in vomit staggering down a road. The studio audience began to hiss with real anger; he looked back in equal anger and said: "Did you all go to high school?" The remark came out faster than his mind, and it was cruel: it went to the heart of the secret. For this audience had a common denominator. Something in their faces spoke of high school as the summit of their lives, the end of their development. These lumpy, dull-necked, dead-eyed people, with their flower-print blouses and acetate shirts, their big-beamed slacks for the women and pinch-cheeked slacks for the men, were living in the kind of torpor that encouraged spending three hours on line to get into a TV show; yes, high school had been the most exciting few years of their lives, and just as they hated him for scoring such an unhappy secret, so did he hate them because he had not had a good time in high school and felt deprived for thirty years thereafter.

Merv Griffin had his work that afternoon. If a studio audience was normally a TV host's prize cow, the cow was acting on this occasion like a bull. The disturbance did not cease, and Mailer, hardly about to hush his views on Vietnam, kept answering the audience's gibes with his own, until he could feel the moment when Griffin must have wondered if the time had come at last when a few dozen people were going to leave their seats and charge onstage. They never did, they never would, but Griffin couldn't be certain that day, and so, for years, whenever he invited Mailer back on his show, he would introduce him as "one of America's best-known brawlers, although he's really a very nice man."

After a time, Dick Cavett began to do something of the same. Once, when Norman suggested he was not a brawler, Cavett gave his fine Venetian smile: "A brawler is a guy who's all marked up after a fight—is that what you object to?"

The audience laughed, and Mailer nodded. Cavett owed him one. Cavett indeed owed him more than one. Once, on Cavett's program, Norman had said he thought David Frost was the best TV interviewer in America, and proceeded to explain to the studio audience that while he had been drunk the night before when he first said this at a taping of the Frost show, the remark had its merits. Carson, he explained, left you feeling like a horse that has been taken through its paces by a no-nonsense jockey, and Griffin, while cordial and decent, was more comfortable with singers and comedians.

"What do I do wrong?" asked Cavett, as if hoping there might still be a joke at the end.

"You're fine."

"No, tell me my flaws. I only invite you on, after all, to give my masochism a fighting chance."

The audience laughed, and Norman was annoyed enough now to pass over Cavett's merits. So he did not say that he thought him the most literate of television hosts, and admired him for trying to keep wit alive in the Sludge (which is the name of the great TV river), but instead said next, "Well, Frost is a good listener." It was true. Frost's gifts were personal. He listened better than other hosts, and so left his guest feeling forceful. "Frost is fun to be on with," Norman concluded.

"I'm not?"

"Oh, I like you."

"I don't care if Frost is better than me," said Cavett, "just as long as you like me." The audience roared.

They moved on to other matters, and in the intermission, Cavett said, "Did you mean it?"

"I said it on Frost's show last night."

"Well, why do you tell me now?"

"This show will go on tonight, but David's won't be heard for another week. I thought it would be unfair to come here, have a good time, and then let you find out later what I said." He paused. "I was going to tell you in the intermission, but it seemed the cheap way to do it."

"Yes, you bet," said Cavett. "David Frost, Jesus Christ!"

The next afternoon, crossing a street in Chicago—Norman was in Chicago to do more TV to promote the sales of a book that no one who watched TV would ever be likely to buy: could it have been *Of a Fire on the Moon*?—he heard a stranger yell out from across the street, "Hey, Norm, how could you ever do that rotten thing last night to Dick Cavett?"

It took years to recognize that he had uttered no little harm; across the seasons, whenever Cavett might have a fight with his network, one or another executive hostile for certain to Cavett's subtle superiority to TV itself would be able to say, "After all, Dick, even your friend

Norman Mailer thinks David Frost is better than you."

Still, he did not come to understand how often the entertainer could have thought about that remark until the night he was on the Cavett show with Janet Flanner and Gore Vidal. It would prove to be an evening when no one would come out well, unless it was Janet Flanner. But it was a night, nonetheless, to convince Mailer that he was incapable of learning how to be a success and get rich on TV.

TO THE EDITOR, *Women's Wear Daily*: It has come to my attention that Gore Vidal has been quoted in your newspaper as speaking with authority of my hatred of women. Let me present the following items.

Number of times married:

Mailer 5 Vidal 0

Number of children:

Mailer 7 Vidal 0

Number of daughters:

Mailer 5 Vidal 0

These statistics of course prove nothing unless it is to suggest that the reason Vidal may have married no lady and fathered no child is due entirely to his love of women and his reluctance therefore to injure their tender flesh with his sharp tongue.*

Yours sincerely,
Norman Mailer

Until this letter, Vidal and Mailer had maintained a professional pact not to speak ill of each other for too little.

Probably this peace had exacted its small price. It is difficult for authors not to say accurate things about

*Since no back copies of this issue of *Women's Wear Daily* are available, the letter is rewritten from memory. It is, however, close, word for word, to the original.

one another in print. Mailer might hear occasionally of a remark Vidal had made in private, but ignore it on the assumption that Gore, tuning up for a little writing, was once again confusing malice with precision.

Still, Mailer knew his own career had become too popular for the pact to continue. Sometime after *The Armies of the Night* had won a couple of prizes, Vidal began to sour in public. Condescending and somewhat righteous references to Mailer began to appear in Vidal's interviews. That was a clue to the temper of the time. Vidal was one literary politician with an impeccable nose for ripe liberal issues, and he was now rushing into the most useful denunciations, politically speaking, against machismo, and was quick to attach to this theme the-hatred-of-Norman-Mailer-for-women as if that were a fact known to the citizenry.

Since Mailer was fond of repeating that there were four stages to comprehending a woman's character, and that you could not claim to know her until you had passed through the four — that is, living together, being married, having children, and going through divorce — since he might therefore like or dislike a lost mate in a given season, but finally thought that was equal to approving or disapproving of Paris or London after you had lived there for many years, a woman indeed being equal to a culture, so that you could hardly pass through four such stages with her and be left with nothing, he thought (as we can see by his letter to *Women's Wear Daily*) that it was gross of Gore to declare that Mailer hated women, when (to be downright crass about it) it was not Vidal, dig! who had spent his life in the collective arms of the powerhouse.

Of course, in compensation, Gore had taken up a seat in the power of print. If people see something in a magazine, they like to believe it. That much Vidal knew. Stupid people! So, sometime after his letter to *Women's Wear Daily*, Mailer picked up *The New York Review of Books* one summer day, July 22, 1971, and read a review by Vidal of a book by an author named Eva Figes, in which the young-old critic stopped long enough to say, "There has been from Henry Miller to Norman Mailer to Charles Manson a logical progression. The Miller-Mailer-Manson man (or M3 for short) has been conditioned to think of women as, at best, breeders of sons; at worst, objects to be poked, humiliated, killed. . . ." Vidal went on from here to speak of M3 as if the case were made—well, in truth, even as Mailer knew that Vidal was looking for just such a result, just so did something blow in Mailer's brain.

"Why," asked Norman Mailer in his mind of Gore Vidal, "didn't you have the simple decency to say, 'This literary progression, extreme as it may sound at first, will engage my best efforts to show that there are startling and frightening similarities between Manson and Miller and Mailer'?"

Instead, this arrogance! Vidal, with his superiority to nuance, would not even make a suitable valet for Henry Miller.

And then, there was himself! He had not been writing for thirty years to be called M3. Mailer could hardly wait to catch up with Gore Vidal.

Yet the opportunity, when it came, did not give good anticipation. Indeed, he hardly knew whether the meeting would take place. He had been invited to appear on Cavett, then he heard Vidal had also been invited, then was told Vidal had dropped off the show. Finally, he was informed Vidal was back again. It seemed Cavett would now have one more guest: Janet Flanner. The assistant producer who offered this piece of news said she was formidable.

Mailer had never met Janet Flanner but was used to seeing her pen name, Genêt, in *The New Yorker.* She had been doing a column on Paris for as long as he could remember (for decades, it seemed, before the real Genet came along), a firm gossipy feuilleton. It was the kind of writing whose merit is not easy to evaluate unless you know as much about the subject as the author; ergo, he had been less impressed with Janet Flanner's literary abilities than by her power to keep a job for a lifetime.

"I assume she's Gore's friend," Mailer said.

"It's more like he's her doll," replied the assistant producer.

"Thank you very much," said Mailer.

It did not bother him. Formidable people were often not formidable under video. Besides, there was no reason to be bothered. By now, Mailer saw no point in being splendid on a TV show—it did not contribute to the sort of meaningful career where every step was on the staircase; in fact, one rarely had an idea of when one was good and when terrible. Mailer had appeared on shows where he felt splendid and full of energy yet looked uncomfortable to himself, even agitated, when he watched it afterward; conversely, he had gone into many a studio feeling lifeless and later looked a model of cool deportment. The secret on TV, if there was a secret, was not to do too much.

Yet it could never be as simple as that. Sitting next to Johnny Carson, for example, was like setting up camp next to an anthill. Carson never stopped fidgeting. None of his nervousness, however, passed through the tube. Finally, one was at the mercy, Mailer expected, of this as yet undiscovered psychoelectronic valve that filtered vices out of certain people only to distort enthusiasm into the appearance of hysteria in others.

Of course, if you could not control how you would appear, if it was up to the valve, then there was no point in getting charged up for a show. Mailer had come to believe in the old Hindu saw: Do not spend time worrying about matters you cannot affect.

So he made other appearances that day. Radio or TV? —he could no longer remember. It seemed from the vantage of Limbo, where old dates were quietly curved (so that you slid in memory around one winter over to another), that he may have been making some kind of effort for *The Prisoner of Sex*—was it for the hard cover or the paperback?—and had, in any case, done at least one show already, and gone to a cocktail party where he had enjoyed three or four drinks to the hilt. Mailer had the operative definition of "to the hilt": it was the state where a carelessly lit match sent you up in flames.

When he arrived at the studio, they rushed him to makeup. Vidal had been supposed to go on first, but hadn't arrived. Would Mailer object to taking his place? He had hardly given his assent before he was informed that Vidal had arrived, after all. Would he now mind if Mr. Vidal went on as originally planned? Since Mailer had formed the little vice when appearing on Cavett of comporting himself as the star, he could not pretend to happiness at finding himself on the shuttle. Still, he kept his mouth shut. He did not wish to jostle his liquor.

At this moment, alone in the greenroom, he felt a tender and caressing hand on the back of his neck. It was Gore. Vidal had never touched him before, but now had the tender smile of a man who would claim, "It doesn't matter, old sport, what we say about each other—it's just pleasant to see an old friend."

Mailer answered with an open-handed tap across the cheek. It was not a slap; neither was it a punch. Just a stiff tap.

To his amazement, Vidal gave him a stiff tap back.

Norman smiled. He leaned forward and looked pleasantly at Gore. He put his hand to the back of Gore's neck. Then he butted him in the head.

"Are you crazy?" asked Vidal.

"Shut up," said Mailer.

"You're absolutely mad. You *are* violent," said Vidal.

"I'll see you on the show."

He was, after Vidal left (and that was quickly enough), obliged to pace about. Other people came into the greenroom, saw him, and went away. It was obvious: he did not feel like speaking.

The show began. Cavett did his monologue, and it was a good one. "A lot of critics knock television. They have a kind of snobbish attitude towards TV, and they think it lacks culture," he said with a smile. He had the only smile that came through the valves of video looking wicked and angelic at once.

"Tonight," continued Cavett, "I have three people on who are important in writing; they've done other things, too, but they're all well-known writers. Just to prove to them that television has figured in the thoughts and writings of people of some stature among the literary profession, I asked my research department to come up with a few things that have been written about this program by some prominent figures. In a couple of cases I have a tendency to do it in the voice of the well-known person. This is from an unpublished work of William F. Buckley Jr. entitled 'A Felicitous Discourse on Epistemological Incunabula,' subtitled 'God: The Man and the Myth.' " The audience applauded his smile with laughter. "The quote is: 'One of the less onerous duties incumbent upon a quondam political journalist is the occasional appearance on sundry quotidian chat programs that proliferate these days like liberals at a hundred-dollar-a-plate dinner for the poor. Not the least in stature, though clearly the least in stature actually, is a jovial dwarf named Cavett' "—they roared—" 'who fortnightly forays into—or nightly, actually—all fields of erudition, and these forays are unblemished by the slightest knowledge of any of them.' That's old Bill." There was laughter.

"Here," Cavett went on, "is a paragraph by women's liberationist Kate Millett . . . entitled 'Little Pig, Big Mouth.' [*Laughter*] 'Let's look at the facts. In the past year, sexist talk-show host Dick Cavett has welcomed several hundred guests on his show. Why have these not included one woman Supreme Court justice, one woman admiral, one woman pro-football player'—he stopped for the laughter—'one woman father, husband, king or rapist? I'll tell you, sisters. Because Mr. Cavett, like all arrogant, ignorant, hypocritical, self-serving, pompous, rotten male-chauvinist pigs, is simply . . .' And then she goes on to become slightly insulting." The audience adored him. "I don't," said Cavett with his finest smile, "want to read that part to you."

A question period with the audience followed, and he was asked, "Do you dye your hair?"

"No," Cavett replied. "I tint my body a little to set it off."

Vidal now came out, and was as sluggish as a club fighter who is looking for his mother.

Vidal: You're looking very well. Do you tint your body? [*Laughter*]
Cavett: Funny, that was going to be my first question to you.
Vidal: That was it exactly, yeah.
Cavett: You heard all that?
Vidal: I was overhearing the audience, yeah.
Cavett: The audience was giving me a bit of a time there, as we say in America.
Vidal: As you say in America. Well . . . love it or leave it.
Cavett: I haven't seen you in a long time. Do they say that in other countries—love it or leave it?
Vidal: Well, yes. Amsterdam, love it or leave it. No, no, you're free to come and go.

Vidal, however, was not a professional for too little. After the commercial, he began to rally; indeed, Mailer was finally obliged to admire how Gore was able (wonder of wonders) to recover from the head butting. Having been taught this informal martial art by a retired black light heavyweight named John Bates who worked for a spell in the late Fifties as a bouncer in the Cafe Riviera on Sheridan Square—Bates's head felt like a cannonball!—Mailer knew the aftermath: nausea and an ongoing vacuum in the brain.

Still, Gore was only showing a few bad effects. Of course, his head had been somewhat harder than expected—writers' heads tend to be more rocklike, for example, than those of actors, who suffer disproportionately from even a tap: head butting may be ultimately a question of how heavy an identity (authentic or false) one is able to support—but, even so, Mailer knew how hard he had hit him. (Between half and three-quarter throttle.) Of course, Gore had given his television routines on so many shows that he could probably recite the same lines, with the same wit, same timing, and same turn of delivery, under sodium pentothal. Moreover, Vidal was not one to forget an effective remark; he never used a new thought so long as the old one could still seem new; it was obvious he had no respect for Mailer's long-held belief that repetition kills the soul. On the contrary, if such a law were in force, Gore would decide it was best not to have a soul. The soul was most famous, after all, for getting in your way—this much cynicism had Vidal absorbed from world literature!

Therefore, the only evidence, at present, that young Gore had passed through a contretemps in the greenroom—that is, had had his head smashed—was his gesture, repeated somewhat frequently, of bringing up several fingertips to massage gently the skin of his forehead, same fingers indeed that had so recently offered to caress the back of Mailer's stiff neck. Other than that, Mr. Vidal was getting around to presenting a first-rate picture of himself.

He was speaking of his friendship with Mrs. Roosevelt. He and she had interests in common, including the protection of the Hudson River from further pollution. For Cavett's liberal audience, there was no name, Mailer decided gloomily, more calculated than Eleanor Roosevelt's to give the wholly satisfactory sensation of having one's buttocks stroked with velvet. To be given intimate glimpses of the lady! Even though he sneered at such crowd-pleasing tactics, Mailer was as taken as anyone in the audience. His only contact, after all, with Mrs. Roosevelt had been at long distance in the Fifties, when he saw her printed reaction to a statement he had made in Lyle Stuart's *Exposé* that white southerners did not wish to give the Negro his equality because—he had been quick to explain—southern whites felt they had their social superiority and Negroes had their sexual superiority, and that—thought southern whites—was fair enough. Eleanor Roosevelt's short statement read: "I think Mr. Mailer's remarks are horrible and unnecessary."

Vidal's life with Mrs. Roosevelt moved, however, on less troubled waters. He spoke of their first meeting with tender éclat. "I drove down," he told Cavett's audience, "to Val-Kill Cottage, where she lived. It was a summer afternoon. The place was absolutely empty. Front door was open, walked in, nobody there. 'Anybody home?' She was a little deaf. No answer. So I went to the—the nearest door was half open—so I opened the door and there, to my horror, was Eleanor Roosevelt—very tall woman—standing in front of the toilet bowl. And I thought, 'Oh, my God.' And she gave a terrible cry, and then she turned around and she said, 'Well, I suppose you'd better know everything.' And she had some gladiolas in the toilet bowl, which she was arranging. She said, 'It keeps them fresh.' " The audience laughed happily. "So we began with perfect intimacy, Mrs. Roosevelt and I. I knew everything from then on that was in the bathroom."

Clever Vidal! Out of the twenty-eight routines suitable for an appearance on Cavett, Gore had chosen to disclose Eleanor Roosevelt's water closet. How could the audience not believe Gore and Eleanor had divine relations?

Vidal (ruminatively): What I liked about her was there was nothing human that she felt human ingenuity wouldn't sense and right action couldn't fix it, and she really believed that, and many people found her very boring because of it. I found her very inspiring. Gosh, I've been with her at four in the morning in Schenectady, trying to get a bus someplace when we were out campaigning, and on and on she would go, this old woman, quite sick, because she felt it was worth doing. And what's gone wrong in the world now—I think a great many of us feel that things are now irreversible, you know. They're never going to make the rivers clear and clean again, they're not going to save the land, they're not going to do anything about Detroit, the American empire is not going to liquidate itself. And there's beginning a sense of despair. You cannot turn it back, and Eleanor was the last person, really, who said, "Oh, we can, and we must."

Janet Flanner now came on. She was not a large woman, but she was old and grand and had an air of giving short measure to fools. By the manner in which she sat down next to Vidal and laid her white-gloved hand on his arm, it was also apparent that she was authoritatively fond of him.

Cavett began by saying that the last time he had had Janet Flanner on his program she had told of finding Ernest Hemingway in her bathtub.

"He was stealing my warm water, that's what he was doing. Sheer theft is what I called it," said Janet Flanner.

She had a nice crust to her personality. It was like the crust on good bread, which lets out agreeable sounds when a piece is broken off. The audience was tittering to show approval. It hardly mattered how much she said, since she could give you a lesson in deportment by the clearing of her throat.

Cavett asked: "Can you bury once and for all the tired old myth of the Frenchman being the sexual, marvelous athlete of all time? . . ." And she replied: "Well, I've buried a few, so I can't tell about all the rest of them." In the midst of the applause, it was now obvious that Cavett, in his turn, was certainly fond of her. His manner seemed to suggest that the only reason he had ever wanted to have his own show in the first place was for the honor of introducing a lady as splendid as herself. He was actually laughing with happiness at her remarks.

For that matter, Mailer was obliged to admit all three were getting on very well.

Vidal: Did you know that George Washington used to take out his false teeth at night and put them in a bowl of Madeira wine?
Cavett: No.
Vidal: I thought you would like to know that, Dick.
Flanner: As they were made of wood, they must have been quite tasty by breakfast. [*Laughter*]
Vidal: It was his brunch that he would put in the next morning.
Cavett: Madeira wine? Why? What's the effect of that on the false teeth?
Vidal: I think he liked the taste, and they were made out of elephant bone and wood.
Flanner: And out of bone? The elephant bone is a new trick to me.
Cavett: I thought they were always wood. I always wondered—I pictured his dentist saying, "I have bad news for you—termites," or something. [*Laughter*]
Flanner: Oh, you're a wicked boy. [*Laughter*]
Cavett: Am I? [*He laughs*] Where are we going?

In the greenroom, Mailer—like that general he could never become—was contemplating the military chances for entering an ambush of such delicacy connected to such strength. The only answer was attack. Shatter all prepared positions. Go out, he said to himself, and smash that fucking teahouse.

THOSE WATCHING would see Norman Mailer emerge onto the screen in a dark and disheveled suit, bow heavily to Janet Flanner, shake hands with Cavett, and pointedly not shake hands with Vidal. People in the studio audience would comment audibly at this display of discourtesy. Mailer would glare at them. Cavett would ask why he did not shake hands. Mailer would reply that he did not approve of Vidal, and thought him shameless in intellectual argument.

The camera would cut to Vidal. He would look composed on occasion, less composed on other occasions. Occasionally, Flanner would touch his arm in reassurance, as if to signify that she would protect him from ruffians. Vidal was obliged to simper at that. It takes real machismo to look stern when your protection is seventy-nine years old.

Under such separate burdens, the next minutes progressed unevenly. Mailer was wound up, and Vidal was silent. Cavett did not seem to know whether to put lanolin on a rash, or pull cocks' feathers. Given this sort of prompting, Norman managed to get in a few words he had written about Vidal in *The Prisoner of Sex.*

Mailer: I said that the need of the magazine reader for a remark he could repeat at dinner was best satisfied by writers with names like Gore Vidal.

Flanner: All those writers called Gore Vidal.
Vidal: I know. There are thousands of them, yeah.
Cavett: Writers with names like Gore Vidal.
Mailer: There are two or three, you see.
Cavett: Who are some of the others?
Mailer: I don't know.
Cavett: Who wants to host the rest of this show for me? [*Laughter*]

Who had butted whom in the head? Mailer now asked himself. After this exchange, his ability to mount a sustained argument, never impressive at best, was hardly robust in the face of Cavett's wit and Flanner's deft interruptions. In short time, the audience turned unmistakably against him. When he said that the contents of Vidal's stomach were "no more interesting than the contents of the stomach of an intellectual cow," they began to boo. He turned to them, and said, "You boo before you know what the man has done." They booed him again. He knew that he must move the conversation over to how Vidal had characterized Miller, Mailer, and Manson as M3, but there was nothing routine about scoring a point when you were debating one on three, or, since Gore was so silent, one on two and a half. During each commercial, Mailer had sat heavily in his seat, staring with a dead eye at Vidal, but this time—airtime certainly racing away—he handed Cavett a page he had torn out of *The New York Review of Books* with Vidal's review, and asked Cavett to pass it over to Vidal to be read aloud. As the show proceeded, however, Vidal did not read it. Norman began to upbraid him further.

"You go in for intellectual adulteration," he said to Vidal, and then, with no great display of coherence, added that Vidal was welcome to rip apart any piece of his writing. "By God," Mailer vowed, "I may be writing on the floor, but if you taught me something about writing, I'd look up and I'd love you for having taught me something about writing. But when you only teach me something about the tricks of adulteration . . ." Mailer shook his head. "You say the rivers will never be clean again, and you pollute the intellectual rivers . . ."

"I wasn't," Vidal drawled, "setting myself up as the Famous Writers School, you know." He drew the chortles he must have anticipated for that.

Mailer: Why don't you try to talk just once, Gore, without yuks? Why not just talk to me insead of talking to the audience?
Vidal: Well, by a curious thing, we have not found ourselves in a friendly neighborhood bar, but both, by election, are sitting here with an audience, so therefore it would be dishonest of us to pretend otherwise. [*Applause*]
Mailer: All right, Gore, look. We are here with an audience by choice, but let us at least—let us at least—

He wanted to grapple with him in argument, get him to read his own words. In the confusion of trying to get to Vidal with one hand and keep Cavett and Flanner off with the other, he was planning to say, "At least read your own words," when Flanner whispered half audibly to Gore.

Mailer: Hey, Miss Flanner. Are you workin' as the referee or as Mr. Vidal's manager? [*Laughter*] I'm perfectly willing to accept you in either role.
Cavett: Can we afford you both talking?
Mailer: But my mind is fragile, and I find it very hard to think, and if you're muttering in the background, it's difficult.
Flanner: I made only the slightest mutter. [*Laughter*] You must be very easily put off center.
Mailer: It's true, you made only the slightest mutter.
Flanner: A tiny mutter.
Mailer: Yes, yes. But I listen to you spellbound—
Flanner: I won't bother you anymore.

Norman was beginning to wonder if anything would work. He made a small vow never to drink again before going on TV. That valuable paper with Vidal's invaluable comments on Manson, Miller, and Mailer was now in Vidal's pocket; oh, that fine review, with such plums as "M3 is on the defensive, shouting names; he thinks that to scream 'dyke' is enough to make the girls burst into tears. . . ." "It was in those years that M3 was born, emigrated to America, killed Indians, killed blacks, conned women. . . ." Miller-Mailer-Manson. Women, beware. Righteous murder stalks the land. . . ."

In the absence of any substance to Mailer's attack, Vidal was taking on strength. The show was past the three-quarter mark, and Gore was coming alive. Yes, Gore was now saying avuncularly, the good thing about Norman "is his constant metamorphosis. He does rebear himself like the phoenix, and what the next reincarnation will be, I don't know."

Mailer: You seem to have figured out that the next reincarnation for me is going to be Charles Manson.
Vidal: Well, you left yourself—
Mailer: Why don't you read what you wrote?
Vidal: You let yourself in for it, and I will tell you—I'll give you a little background here—that Mailer has—
Mailer: We all know that I stabbed my wife years ago, we do know that, Gore. You were playing on that.
Vidal: Let's just forget about it.
Mailer: You don't want to forget about it. You're a liar and a hypocrite. You were playing on it.
Vidal: But that wasn't a lie or a hypocrisy.
Mailer: The fact of the matter is that people who read *The New York Review of Books* know perfectly well—they know all about it, and it's your subtle little way of doing it . . .
Vidal: Oh, I'm beginning to see what bothers you now. Okay, I'm getting the point.
Mailer: Are you ready to apologize?
Vidal: I would apologize if—if it hurts your feelings, of course I would.
Mailer: No, it hurts my sense of intellectual pollution.
Vidal: Well, I must say, as an expert, you should know about such things. [*Laughter*]
Mailer: Yes, well, I've had to smell your works from time to time, and that has helped me to become an expert on intellectual pollution, yes.
Vidal: Yeah, well . . . let's—I was going to say, I—
Flanner: Not only do you insult each other, not only in public, but you act as if you were in private. That's the odd way—

Mailer: It's the art of television, isn't it?

Flanner: It's very odd that you act so—you act as if you were the only people here.

Mailer: Aren't we?

Flanner: They're here, he's here, *I'm* here, and I'm becoming very, very bored. [*Laughter, applause*]

Mailer: You still haven't told me whether you're Gore's manager or the referee.

Cavett: If you make history here by punching a lady . . . [*Laughter*]

Flanner: I won't have it! I won't have it!

Mailer: Now, look, you see the sort of thing that goes on. Now you say I make history by punching a lady. You know perfectly well—you know perfectly well that I'm the gentlest of the four people here. [*Laughter*]

Cavett: I just hope it lasts through the next whatever we have left. [*Laughter*]

Mailer: I guarantee you I wouldn't hit any of the people here, because they're smaller.

Cavett: In what ways? [*Laughter*]

Mailer: Intellectually smaller.

Cavett: Let me turn my chair and join these three. [*Laughter, applause as he moves his place*] Perhaps you'd like two more chairs to contain your giant intellect. [*Laughter, applause*]

Mailer: I'll take the two chairs if you will all accept finger bowls.

This remark was sufficiently gnomic for Cavett to chew and get to no witty place.

Cavett: Who wants to grab this on our team? [*Laughter*] I nearly have it, it means something to me. Finger bowls. Finger bowls. Things you dip your fingers in after you've gotten them filthy from eating. Am I on the right track? Am I warm?
Mailer: Why don't you look at your question sheet and ask a question.
Cavett: Why don't you fold it five ways and put it where the moon don't shine.

It was received as the remark of the evening. Borborygmic was the laughter. Mailer sat there, contemplating a region in the depth of his rectum, "Mr. Cavett," he said, "on your word of honor, did you just make that up, or have you had it canned for years, and were you waiting for the best moment to use it?"

"I have to tell you a quote from Tolstoy?" said Cavett, and the laughter continued.

Mailer turned his chair away from the guests and toward the audience. "I want to ask all of you something," he said to the audience. "Are you all really, truly idiots, or is it me?"

The audience replied: *"You!"*

"Oh, that was an easy answer," said Cavett.

The music came up. Another intermission was upon them.

Cavett: I just wonder what the people at home think happened during the break. The four of us sat here in stony silence and—

Vidal: No, somebody in the audience shouted, "Your argument is immature," to all four of us.

Cavett: To all four of us. Oh. So we can each take as much—

Flanner: We can each take a bow.

Cavett:—as we feel we deserve. I'm sorry, Norman; I did interrupt you, and you were talking to the assembled audience.

Mailer: Yes, I was going to ask the audience what I was doing that was making them cheer every time the other side connected with a pass from one to the other of my three opponents.

Man in audience: You're rude.

Woman in audience: You're a snot.

Mailer: That's fair. Someone said I'm rude and someone said I'm being a snot.

Woman in audience: You're a pig.

Mailer: Ohhh, the joint is loaded with libbies, ohhh. [*Applause*] Gore, my God, it wasn't enough to trundle Janet Flanner along, the most formidable presence in the history of television, but you had to load the balcony with your libbies, your little libbies.

Cavett: Well, now, wait a minute. There was only one lady's voice that rang out. Do you consider that loaded?

Mailer: It was the voice of legions.

Woman in audience: Why do you have to argue so negatively and insultingly to your guests?

Mailer: They're not my guests. They're not my guests any more than they're your guests.

Cavett: It seems it's your show now.

Woman in audience: Why do you have to answer them

with insults and nasty statements, and they're answering you maturely and with dignity? [*Applause*]
Mailer: That's because they're mature and full of dignity and they'd cut my throat in any alley, and I answer rudely because I'm crude and a lout and a clod, that's why. [*Applause*] Now, have we progressed at all? Can I reach you, can I talk to any of you, or is it hopeless? I mean, if you care to listen, I do have a thing or two to say, believe it or not.
Man in audience: Why are you bellyaching so much?
Mailer: Well, I'm bored with being misrepresented, you see. I've been misrepresented, by my own paranoid lights, for twenty-five years in this country, something like that, and to find an old litterateur like Gore Vidal, who knows his way around, you know . . . knows the ropes, and then suddenly, after all these years, he's suddenly developed a particularly dirty literary game. Well, it inflames me, you know. I've been so bold as to pretend to be the presumptive literary champ, you know, whether I deserve to be or not. The reason people always talk about me in relation to Hemingway is just that Hemingway at a certain point said to himself with his huge paranoia, "They're going to kill me for this, but I'm going to be the champ—it's all I care about." And he shifted the course of American letters because up to that point people who wrote books were men of letters, they were gentlemen, they wrote books, and Hemingway said, in effect, "No, people who write books take as much punishment as prizefighters, and one of them has to be a champion." Now, we've had a time that has not been interesting in terms of champions. In

Hemingway's time there were great writers: there was Faulkner, there was Hemingway, and there were many other writers, like Steinbeck and Farrell and Dos Passos and Thomas Wolfe, many, many really wonderful writers. Our time has been much more complicated, and there haven't been that many really extraordinary writers around, and I have presumed, with all my extraordinary arrogance and loutishness and crudeness, to step forth and say, "I'm going to be the champ until one of you knocks me off." Well, fine, but, you know, they don't knock me off because they're too damned simply yellow, and they kick me in the nuts, and I don't like it.

That was perhaps the most impassioned speech he had ever delivered on television. It said little, but it was full of feeling, and it contained the cardinal virtue of television (for any ambitious performer), which was that it kept attention on himself. It suggested that his close-up in the scheme of things would demand more minutes than would those of the others. The audience, obedient to these conditioned reflexes of television, actually gave him a hand, a powerful hand of applause. But, of course, he was hardly finished.

The time had now come to give his peroration on women's liberation. He was not opposed to it, he wished to tell the world; he even honored it. It would free the best of women to discover how brave they were; a world in which women were maintained systematically in cowardice had to be a bad world. But, look, he wanted to say, as in all revolutions, the worst are running with the best, and there is an evil loose in women's liberation.

It is the idea that the male search for bravery is mean and ridiculous, and that it is easier to be a man than a woman. Destroy the difference in the sexes, he would tell this audience, and a world of asexual human units, ready to fit the computers of a future totalitarianism, would also be prepared; these were the questions Vidal was polluting with his use of M3.

The speech might have been worth giving, but the commercial came before he was half launched, and what he had to say was unsaid.

During the intermission, they argued on how to proceed. Vidal now wanted time to reply. Cavett indicated that it was Cavett's place to decide; he decided Vidal was entitled to equal time. Mailer offered to accept if Vidal would at least read from his review. With nothing agreed, the music came up.

Cavett: Gore, the aim narrowed down to you, and we have three and a half minutes left. I insist that you get them.

Vidal: Well, I'll begin to answer Norman's charge about what a bad person I am. The attack on him, really, if you want to know, Norman, is simply what I detest in you—and I like many things in you, as you know. I'm a constant friend despite this—but your violence, your love of murder, your celebration of rage, of hate. . . . *American Dream*—what was the dream? A man murders his wife and then buggers this woman afterwards to celebrate an American man's dream. This violence, this knocking people down, this carrying on, is a terrible thing. Now, it may make you a great artist—

Mailer: I demand one minute at the end—
Vidal: It may make you a great artist—
Mailer: I'll listen, but I demand one minute at the end—
Vidal: It may make you an interesting artist, I don't say that, but to the extent that one is interested in the way the society is going, there is quite enough of this stress, quite enough of this violence, without what I think are your celebrations of it, your attitude towards women in this thing, which I thought really horrible, and you said I compared you to Charles Manson. I said Henry Miller in his way, Norman in his, and Manson in his far-out mad way are each reflecting a hatred of women and a hatred of place.

One person was applauding loudly. At that moment, Norman got up from his chair, took a few steps over to Vidal's chair, flipped the review from *The New York Review of Books* out of his hand (for Vidal had been waving it as he spoke), and then returned to his seat. Vidal had been in the thick of saying, "And frankly, if I may say so," when he saw Norman's bulk looming above, and he flinched. Vidal flinched discernibly: a monumentally high number of electrons were thereby shifted in millions of TV sets.

Back in his seat, Mailer began to read from Gore Vidal's review.

Mailer: "There has been from Henry Miller to Norman Mailer to Charles Manson a logical progression." Period. "The Miller-Mailer-Manson man (or M3 for short) has been conditioned to think of women as, at best, breeders of sons, at worst, objects to be poked, humiliated, killed." And from there on in the piece you speak of Miller, the great writer Henry Miller, the greatest writer alive in America—if we're going to talk like muckers, I'll talk, too, like a mucker—Henry Miller, the greatest writer alive in America, and myself, and Charles Manson, a hugely complex and contradictory figure, are spoken of, lumped together, as M3, and if you call that good intellect working, to lump together three people as curious as Henry Miller, Norman Mailer, and Charles Manson...

Vidal: Well, you must read the piece. You can't—

Mailer: I read it.

Vidal: You happened to read it, but the audience has not. You are selecting this one passage as representative of the whole. I made my case very carefully. But I will say, giving you a few minutes more on the program, you will prove my point. So, it is—I come back to what I said. I detest this violence in you. You have actually written that "murder is never nonsexual."

Mailer: Well, is it ever nonsexual?

Vidal: Well, I'm—

Mailer: Don't you know, Gore?

Vidal: Not having murdered anybody lately, no, I don't know.

Mailer: You bragged about what you did to Jack Kerouac, after all.

Vidal: He didn't die.

Mailer: Well, he did.

Vidal: Oh, come on.

Mailer: Don't you remember the stories you used to tell about him?

Vidal: I'm going to give you a line that Degas said to Whistler—two celebrated painters—and Whistler was a great performer, like Norman, and Degas said, "You know, Whistler, you act as if you had no talent." You represent yourself as though you really had no talent at all, and of course you are one of our best writers.
Mailer: I read it in the same place you read it, which is in Edmund Wilson's answer to Vladimir Nabokov in last week's *Times.*
Vidal: So what? What's that got to do with it?
Mailer: I thought it was marvelous.
Vidal: Good. I'm happy that we both agree that the sentiment is correct.
Mailer: The difference is that I savor the remark and you throw it into the battle.
Vidal: Oh, Norman.
Mailer: Now, may I have my one minute for rebuttal?
Cavett: There is one minute remaining.
Mailer: There is one minute. All right. He speaks of a character in a book of mine murdering his wife and then buggering a woman as a celebration of American manhood. Now, I throw myself upon your court, hanging judge. I say that's intellectual muckerism of the worst sort because the character I have there was a particularly complex character, and, in fact, he did not simply bugger a woman, he entered her the other way as well, and there was a particularly complicated—
Flanner: Oh, goodness' sake. [*Laughter*]
Mailer: I know you've lived in France for many years, but believe me, Janet, it's possible to enter a woman another way as well.

Flanner: So I've heard. [*Laughter*]
Cavett: On that classy note . . . [*Laughter*]
Flanner: I don't think it's restricted to French information, dear.
Cavett: She said she doesn't think that's restricted to the French.
Flanner: Practically international.
Mailer: Are we quitting on that classy note?
Cavett: On that classy note, we have a brief message from our local stations, and we will be back.

Cavett and Mailer were not looking at each other. Vidal was looking off into space. Janet Flanner was grinning at Mailer as if he were a most curious fellow. What a way to use his last minute!

The music came up.

Cavett: Well, this has been an interesting evening around the old table, and, Miss Flanner, I'm glad to see you got that box of cookies that you wanted. And you other two fellows—
Flanner: It's the only solace.
Cavett: Yeah. [*Laughter*] Could you two come back New Year's Eve, and maybe we can—[*Laughter*]
Mailer: That's the night.
Cavett: Christmas Eve. Well, let us know who you think won. That'll be interesting, and we'll see you next time we're on the air. If. [*More laughter and applause*]

The good-byes were short. Mailer turned around, and Vidal was gone.

AFTERWARD, long after the mail stopped coming (which took a time in itself), after the magazine accounts were finished and he no longer heard students ask, "Were you really mad at Mr. Vidal?" every time he gave a lecture, long after he had lost every last bit of interest in the evening and remembered the program only because accounts kept coming back to him of Vidal saying on one network that Mailer-is-finished-as-a-writer and Norman-is-bad-news on another, a string of claims that was now in its sixth year (and would never end, since he knew he would never run into Vidal for an accounting—once, when Norman was in Rome for a month to work on a movie script, Vidal managed to leave for New York), he finally had time to recognize that this Dick Cavett show, if it was one of the few occasions when television had come alive, had also been killing to the last of his hopes that television might be a forum for those curious ideas with which he lived.

Hundreds of letters came in, ten times more, or maybe it was fifty times more, letters than he had ever received from any other show, and as many, he was certain, went to Vidal—all those lovers of Eleanor Roosevelt! His mail was favorable to him, even as Vidal's mail must in turn have been favorable, but the letters, if they brought a sense of his real audience, were not in response to any triumph he had had over the medium, but were more a note, even a chorus, of outrage, at the restraints of television, the villainy of Cavett in choosing sides, the inanity of the studio audience, and much on Vidal's little flinch—hardly a letter did not pride itself on seeing that Vidal had flinched.

In turn he hardly had to speculate on how other letters must be describing him: a boor, a drunk, an incoherent and unworthy opponent for Mr. Vidal.

It had been Bullshit Bisque when all was said. He had not only failed to dramatize his ideas, but had dramatized himself—it was himself rather than any idea that had come through the tube. He had succeeded, at best, in evoking sympathy for his own lack of taste and polish; yes, when it was all done, and psychic horizons could be glimpsed again, he knew that he had wasted another fierce piece of time, and Limbo was by that measure more advanced on him, even knew at last from the vantage of Limbo, from the intimacy of this observation into the interstices of his past, that it did not matter on how many TV shows he would yet appear, on Cavett again and Carson and Griffin and Douglas and Susskind and *Today* and *Tomorrow* and *60 Minutes*, no past and no future, no antipathy and no forgiveness, no, it did not matter, he had crossed over to the other side on this night of the Dick Cavett show, and now saw that television was an addiction whose pleasure (since every addiction has its pleasure) was to give a small sense of manners to the mannerless and to pass a hint of style over to those who are hungry with some unfulfilled vision of themselves (even as he had once studied TV). Yet, it would exact, like any drug, its own cruel price. That took many a form, but for our purposes the price may be equal to the power to destroy the virility and fecundity of ideas. TV was not with us to make history, but to leach the salts of history right out of our cells.

Since Mailer came to know this in his life, and yet continued to go on TV, he advanced from a wicked man to the beginning of an evil figure, at least to his own eyes, which here were unhappily congruent with the eyes of Limbo. If it was wicked to up the ante, it was evil to come to know the good and defy it, and the good, in this case, was to keep the hell off TV, which he did not. So he came to know his place in Limbo before he was even there.

That is our story and we could be done if it was not demoralizing to cease at such a point, blanked in Limbo and empty of hope.

Since it is also the work of such a place, however, to cleanse the choked-up rivers of our lives and thereby enable us to voyage out to a more adventurous plane (where the walls are rough and the ascents are vertical), not every meditation in the mansion need brood over the hopelessness of the past. There is also the memory of odd and mysterious instants when hope is in the air, even if comprehension is dim. So Mailer, sentenced by Limbo to remember no experiences that were altogether separate from television, was nonetheless able to think fondly of two events he saw from the spectator's side of the tube, and was agreeably overcome with the suspicion—suspicion proving tonic to the flattened recognitions of Limbo—that even in the middle of *something* there might be a force devoted to the extinction of *something*. If the Devil always appeared in the most splendid intentions of the Lord, the Devil being the most beautiful creature God ever made (in those days the Lord was young and still vain in His vision and therefore wicked enough to create beauty before prudence), so, too, had God been obliged in His turn to enter all those baleful and satanic mediums where the air is destroyed and electromagnetism chases its tail and shatters frequently; yes, maybe God was now as grim as the glimpse of defeat in the center of television when the set goes off, and yet was also ready, though no longer young, to chase the Devil into every void, live in every static, and go to war on each bleak field. So there might be hope. Mailer did not know—Limbo gave no purchase on the eschatological structure of things, but still!

Mailer remembered the night he invited friends to his house in Provincetown back in the summer of '66 because his great friend José Torres was defending the light-heavyweight championship against Eddie Cotton in Las Vegas, and the fight was on TV. That was an occasion to remember in Limbo, and to see again in Limbo, yes.

Torres was a strong favorite that night, but Cotton gave him trouble. Eddie Cotton was an old black fighter with a thousand tricks and more than a hundred fights, and being near the end of a distinguished and rugged career, he was ready to steal the championship if he could put it all together, and Cotton put it together. He fought the fight of his life. Before half of the fifteen rounds were gone, he was clearly ahead of José Torres.

There were eight guests in Mailer's living room, and they were quietly drinking and commenting on the action. About this time, Norman began to go out of his mind. He did not really know a lot about boxing, not so much as he was reputed to know, but he knew Torres' style and had an insight into Torres' strengths, and on this night he felt as if he knew what Torres was not doing right. "Stick him," he began to shout at the set. "Get your jab working." Torres' vice, if he had one as a fighter, was vanity, and Torres was being beaten on combinations by Cotton. Since José's vanity was that nobody was better at combinations than Torres, he had been trading his best combinations against Cotton, and for one round after another, Cotton had been beating him by just a little. What a fight! One was seeing some of the finest combinations ever to come out of a war between light heavyweights, fantastic combinations indeed for light heavyweights. It was a great fight, but Torres was not using his jab, that jab off his long arm that had been finer even than the jab of Willie Pastrano on the night Torres took the championship from him, and so Norman began to scream at the set in a purple and apoplectic voice, "Stick him! Get it going! Stick him!" And as if Mailer had been in Torres' corner and Torres were listening to him, so Torres began to pop Cotton with his jab; there in the middle of the fight, Torres shifted his pattern, and began to jab, and this threw Cotton off just a hair on his combinations, and the jab began to take the edge from Cotton and to give it back to Torres, and Norman, watching the action, swallowed bourbon like gasoline, and screamed instructions—"Keep it up! Keep it up! Keep the pressure!"—to what must have been the peculiar horror of his guests; now more rounds were going to Torres than to Cotton, and Norman screamed him home, eight rounds to seven, and José kept his championship. Norman was never to get over the impression that somewhere in the depths of Torres' attention the message had come through the tube: "Stick him!" A miracle in communication had occurred. A TV set, that malign corporate mechanism of valves that allow communication to pass in one direction only, had been transcended for half an hour and managed to take messages back, or to say it real, the heart of Mailer's insanity that night was to think he had passed through the heart of *something* and was talking successfully to the tube.

Insanity has its rewards. Later that night, convinced he had broken through some encircling ring of the Devil, and was now out of his box, that is to say alive in his life rather than sequestered in a psychic coffin, Norman began to run amok at a dress rehearsal of his play *The Deer Park*, which would have its premiere the following night at the Art Association in Provincetown. He had kept his mouth shut at rehearsals when he wished to interfere; he had allowed, he thought, his director to take *The Deer Park* down lanes the play should not travel, and on this night, after the Torres-Cotton fight, having come to the rehearsal late, and lately filled with a sense of power few humans could know—he could talk back to the tube!—he saw no reason why he should not talk back to his actors, and did. Sitting at the back of the hall, hardly able to hear their voices, he began to shout, "Louder!"

"Norman, will you please be quiet?" said the director.

"I will not be quiet. I cannot hear the actors."

They had a quarrel in front of the cast. It grew worse,

and the director, in a rage, called the rehearsal off. It was only on the next day that Mailer found out he had been witnessing not a dress rehearsal, but a walk-through for light cues, and that was why he could not hear the actors. They were saving their voices.

"When it comes to theater, you are an ignorant man," said the director, Leo Garen, the next day, and since he was a young director and had no money at the time, and this play, if it went on to Off Broadway, was his best prospect, Mailer knew the courage it required for Garen to tell him off, and shook his hand on the spot—it was somehow relaxing to hear, in the middle of this fierce hangover, that he was ignorant; he had not been called ignorant in many a year; some insults come like a blow on the head the morning after, but a few are balm—and he later hired Garen to take *The Deer Park* to New York, and thus Mailer began a short career in theater and soon invested his own money in making and directing three films, and by way of them proceeded eventually to lose all his money and more, and so was obliged to keep writing fast for the rest of his life, a blessing to those who liked his stuff, a curse to those who didn't, but much change had certainly come to his life by daring to go up through the tube.

If this pale reminiscence of the rehearsal of *The Deer Park* was as far as he had been able to stray in Limbo away from grim recollections of television, still it cheered him, even if the new images were still of himself as a fool; at least they were recollections free of video. If such memories continued, he might even recall a happy copulation or two. Of course, he did not have the hope that Limbo would permit any rounded view of one's powers of compassion and detachment, fear and greed, strength and responsibility, as a good friend and a bad one, a lover and a monster of love, as egocentric and selfless; no, Limbo was strict, Limbo repaid you for dabbling in television by making you live in it; these walls, like the inside of a TV tube, were equal to the insomnia that got you up at two in the morning to peer into the box. He had never expected that the scale of payment would be as intimate as your worst enemy's judgment on your hopes.

YET, as his mind went on, and as he came to retrace every episode on this avenue of his sins, an itinerary for which he did not have to move a step, since the curved walls of Limbo served splendidly as a screen, he was at last entitled to move out of his most intolerable punishment (which here for the narcissist, paradoxically, is constant and enforced contemplation of the self) and to begin to consider the merits of others.

So he came at last to the one night of TV that had given him real pleasure, and he thought often about it. In his partial comprehension of television, in his researches into the mystery of that hum in its void, he had always centered his hypotheses around the familiar certainty that Richard Nixon owned the human personality closest to the persona of the television set. Mailer had even built his distrust of TV and of Nixon on their similarity to each other, their incorporation of baleful disregard for the finer possibilities of human attention.

Yet, when he replayed in his mind, that is, on the curved walls of the mansion, Nixon's first appearance before David Frost, in which the ex-President spoke of Watergate, Mailer was cheered by that ninety minutes. In recent years, his detestation of TV close to complete, ninety minutes had become an immense dose of video, and left Mailer feeling like a canceroid (that is, like someone who could not justify why he should not bear the disease). Yet Nixon's ninety minutes were of another order; on the scale of video, they were alchemy itself: the American with the least charismatic personality of the century, the American most ready to be emitted from the tube and infiltrated into your pores, had become for a few minutes a personality capable of moving multitudes, and had done it by the refractory art of the actor.

He had always believed Richard Nixon was the most untalented actor he had ever witnessed, and Mailer had often pondered the first meeting decades ago of Richard and Pat Nixon in an amateur theatrical company, and had wondered whether Nixon, if he had been somewhat better in those early years as an actor, would have become in later life a politician more like Ronald Reagan. Instead, Nixon had become one of the few politicians on the American scene who was a consummately abominable actor. His lack of conviction vibrated through the diodes, triodes, and, later, the transistors of every American set; the message he gave was profoundly depressing. Even as a district attorney who speaks in too righteous tones to a jury will depress us on questions of law and order and their lack of relation to justice (since the fact that a D.A. can prosper on such hypocritical terms has either to speak ill of the jury or of a system that encourages us to believe that phony voices carry the real tones of avenging authority), so Richard Nixon's inability to act had always intimated that something in the American public must be atrocious if millions were ready to accept his transparent lack of sincerity, his push-button smile, and his simple lack of ability to offer even that resonance of the throat which is the ham actor's emotion.

Yet, in the interview with Frost that Mailer witnessed, a miracle was visible—Richard Nixon had become not a good actor but a great actor, great by the measure of actors like Bogart and Fonda and old Spencer Tracy, great—good God!—like Edward G. Robinson, like Edward G. Robinson playing Richard Nixon full of woe at what had come to him from Watergate, and after thirty or forty minutes of watching Nixon, it did not matter what the truth might be, any more than one would find fault with a great stage actor for bringing life and splendor and passion and the monumental echoes of tragic woe to lines that were not his own and that he could say in his sleep or while shaving, he had practiced them so well; yes, Nixon struck America with a miracle—a talentless actor had become a splendid actor—yes, Nixon now went to the root of good acting, where before he had lived in the center of bad acting. In his political years, Nixon used to comport himself as if acting were a set of semaphores and you frowned if you wished to look stern, and glanced heavenward to demonstrate that your motives were divine; now, face-to-face with Frost, he was an actor from the guts out, which is to say that he got his words out through every effort to control himself. The great actor does not play a drunk by staggering; rather, he plays a man whose stomach is raw and whose head is whirling, but a man who nonetheless believes he is sober and so will make every effort to convince others he is sober. The impression of drunkenness comes over the audience by virtue of the actor's attempts to conceal drunkenness. That was Nixon's power on this night. What a mighty emotion he was seeking to hold like a man! It did not matter after a while that a part of one's brain could still remember many an intricacy of Watergate that Nixon was debauching once again; that was like saying of Olivier in *Marathon Man* that he might be playing an ex-Nazi but actually was still Olivier and born British; what indeed did the truth of Nixon's guilt have to do with the higher truth that a talentless man had become a powerful actor? A fact such as this was kin to a miracle and warmed the heart. So as Richard came forth with those beautiful and unforgettable lines, beautiful, that is, and unforgettable as examples of the actor's, not the playwright's, art, as he said: ". . . it was springtime. The tulips had just come out . . . it was one of those gorgeous days when, you know, no clouds were on the mountain. And"—his voice went husky here from the effort not to be husky—"I was pretty emotionally wrought up, and I remember that I could just hardly bring myself to tell Ehrlichman that he had to go. . . . I said, 'I hoped, I almost prayed, I wouldn't wake up this morning.' Well, it was an emotional moment. I think there were tears in our eyes, both of us. He said, 'Don't say that.' We went back in." Then he said goodbye to Haldeman and Ehrlichman, and took their resignations. ". . . It was late, but I did it." No actor alive, not Ralph Richardson, Jack Nicholson, or John Gielgud, could have given a better reading than he gave on the next line. "I cut off one arm," Nixon said manfully, and in partial control of great pain, "and then cut off the other arm." He nodded. ". . . You could sum it all up," he said to Frost, "the way one of your British prime ministers summed it up, Gladstone, when he said that 'the first requirement for a prime minister is to be a good butcher.' Well, I think . . . I did some of the big things rather well. But I . . . have to admit, I wasn't a good butcher."

(It would come out later that Nixon's version of these events was not necessarily accurate; first Ehrlichman, then Haldeman, would dispute what he had said—he was not cutting off his own arms so much, it seemed, as he was cutting off their legs.)

Yet with what a difference did he lie! In the past, a deadening actor and a lifeless politician, Nixon had lied with every dull light in his eye; now his eye was bright, he had learned to act: his lies partook of art.

The impact upon Limbo was immense. Limbo had many mansions and took many shapes, but on the rock that was its foundation were the words: Men like Richard Nixon are not redeemable.

Nixon had shattered the rock. Reflecting on Nixon's Ode to the Prime Minister and the Art of Butchery, Mailer felt the breaking of the rock. Nixon had demonstrated that if one's suffering were great enough, even a hopeless actor could learn to act, and that meant millions of Americans dragged into the pit by Nixon and forced to believe that bad acting, transparent hypocrisy, is the main road to success would now be obliged to stare into the true pit: *Life demanded that one improve.* Even a hypocrite was wise to sin with a light in his eye. So the fundament of Limbo cracked. It could not decide whether Richard Nixon was still the avatar of video, a TV set with a head and limbs, and video itself was altering into a higher form of the malignant; or whether Nixon, having betrayed all other gods, had now betrayed the Devil as well, and would try to make a last career for himself good enough to bring Broadway back.

Mailer did not have time to care. He knew only that Limbo was looser now, and he was getting ready to move from the mansion; to go on—where to he did not know: a curiosity or a horror could await him next. He hoped only that tomorrow—in the long tomorrow of Limbo—would lead to autumnal places where the record of his sins might reek of woodsmoke rather than of the insulation of all those TV wires charring the night.

It was on this note that he was allowed to think of a poem. The irony was that it was by a poet, James Elroy Flecker, he could hardly have read before. Only the last two lines would stay in his mind. "Oh, is it," said Mailer to himself,

Is it the mist or the dead leaves?
Or is it the dead men—November eves?

THIS first edition of OF A SMALL AND MODEST MALIGNANCY, WICKED AND BRISTLING WITH DOTS is limited to three hundred numbered copies on Mohawk Superfine and a deluxe edition of 100 copies on Curtis Rag bound in leather, all of which have been signed by the author.
The type is Hermann Zapf's Aldus.
Binding by Bela Blau.
Designed by Grant Dahlstrom and printed at The Castle Press for the Lord John Press.

This is number 73